Planetary Habitability

AMERICAN ASTRONOMICAL SOCIETY IOP | ebooks™

AAS Editor in Chief

About the program:

AAS-IOP Astronomy ebooks is the official book program of the American Astronomical Society (AAS) and aims to share in depth the most fascinating areas of astronomy, astrophysics, solar physics, and planetary science. The program includes publications in the following topics:

- GALAXIES AND COSMOLOGY
- INTERSTELLAR MATTER AND THE LOCAL UNIVERSE
- STARS AND STELLAR PHYSICS
- EDUCATION, OUTREACH, AND HERITAGE
- HIGH-ENERGY PHENOMENA AND FUNDAMENTAL PHYSICS
- THE SUN AND THE HELIOSPHERE
- THE SOLAR SYSTEM, EXOPLANETS, AND ASTROBIOLOGY
- LABORATORY ASTROPHYSICS, INSTRUMENTATION, SOFTWARE, AND DATA

Books in the program range in level from short introductory texts on fast-moving areas, graduate and upper-level undergraduate textbooks, research monographs, and practical handbooks.

For a complete list of published and forthcoming titles, please visit iopscience.org/books/aas.

About the American Astronomical Society

The American Astronomical Society (aas.org), established 1899, is the major organization of professional astronomers in North America. The membership (~7,000) also includes physicists, mathematicians, geologists, engineers, and others whose research interests lie within the broad spectrum of subjects now comprising the contemporary astronomical sciences. The mission of the Society is to enhance and share humanity's scientific understanding of the universe.

Planetary Habitability

Stephen R Kane
Department of Earth and Planetary Sciences,
University of California, Riverside, USA

IOP Publishing, Bristol, UK

ISBN 978-0-7503-2120-4 (ebook)
ISBN 978-0-7503-2118-1 (print)
ISBN 978-0-7503-2121-1 (myPrint)
ISBN 978-0-7503-2119-8 (mobi)

DOI 10.1088/2514-3433/ac2aa1

Version: 20211201

AAS–IOP Astronomy
ISSN 2514-3433 (online)
ISSN 2515-141X (print)

British Library Cataloguing-in-Publication Data: A catalogue record for this book is available from the British Library.

Published by IOP Publishing, wholly owned by The Institute of Physics, London

IOP Publishing, Temple Circus, Temple Way, Bristol, BS1 6HG, UK

US Office: IOP Publishing, Inc., 190 North Independence Mall West, Suite 601, Philadelphia, PA 19106, USA

Cover image: Courtesy of Don Brownlee, the University of Washington

Contents

Preface

This book is incomplete. The topic of habitability is so complex that it is near impossible to be all inclusive on the matter, and there are undoubtedly issues that are not covered completely in this book. When I was asked to write a book that summarizes planetary habitability, my immediate response was, "Is now the right time?" The various disciplines that contribute to the habitability conversation are advancing at a spectacular rate, and discoveries are occurring at a pace that is truly breath-taking. In the midst of all this, I realized that, although there is still much work to do, waiting for the discovery dust to settle is at best an asymptotic exercise. As such, I needed to be okay with this book being incomplete before I began.

Another aspect I considered carefully in approaching this book is the intended audience. As noted above, planetary habitability is a diverse subject that relies on the expertise of many people across a range of disciplines. Indeed, the students who have taken my graduate class on this topic have had backgrounds that include astronomy, biology, climate science, geophysics, and planetary science. As such, I chose to write this book for that demographic in a way that maximizes the possible entry points while teaching the necessary fundamentals, and providing direction to other resources should they be needed. My intent is that anyone from these discilpines can pick up this book and obtain a broad understanding of the larger picture, and spur ideas on how their particular research might contribute to the habitability story.

When it is inevitably pointed out that this book doesn't include process X, I will probably nod my head and agree. However, it is exactly such discussions regarding what is and is not important for the habitability issue that are rapidly pushing the field forward. As we are developing a deeper appreciation of the inter-disciplinary and system science approaches that are required, we are gradually realizing the vast depth of the habitability subject. The data collected for the solar system bodies have never been more important for assessing habitability, as the exoplanet community prepares to apply the many lessons learned over the coming decades, to objects that we (for the most part) cannot even directly observe. Therefore, inferences of habitable environments that are many parsecs away and for which we will have no *in situ* data will require a high level of confidence in our models based on solar system observations. My hope is that we are now laying the foundations for further discussions that will continue to disentangle the plethora of influences on overall planetary habitability.

Stephen R. Kane
Riverside, July 2021

Acknowledgements

On my journey through the topic of planetary habitability over the past decade or so, I have had the incredible fortune to work with amazing students and colleagues. If I listed everyone with whom I have spoken about this subject, then I fear the acknowledgments would dominate the book. Here I restrict myself to mentioning a few people who have had a major influence on my work, with the recognition that this work really has been the result of interactions with hundreds of colleagues over many years.

First, I would like to thank my friend and colleague Ravi Kopparapu for being such an incredible sounding board against which to test my ideas. I remember when we first met in 2011, in the restaurant at Extreme Solar Systems II in Jackson Hole right before the meeting started, and it has been my honor and privilege working with you ever since.

To Dawn Gelino, my partner-in-crime in developing and launching the Habitable Zone Gallery, which saw "first light" back in August, 2011. Thank you for your constant support of my work as I rapidly diverged from exoplanets into planetary science, habitability, and beyond. Thanks also to Colin Chandler and Natalie Hinkel for their help in improving the Habitable Zone Gallery and for many useful discussions.

I am grateful to Elizabeth Tasker for her inspirational writing ability, of which I hope to one day achieve a small fraction of that capability. Thank you for all of your valuable wisdom and advice in approaching this work. Thanks also to Maggie Turnbull for many thought-provoking conversations regarding habitability and direct imaging prospects.

I offer my heartfelt thanks to my Venus colleagues, who welcomed me into their community almost a decade ago, and have listened to all my rants about the Venus/Earth habitability divergence. I would like to especially thank Giada Arney, Paul Byrne, Bradley Foley, Jim Head, Noam Izenberg, Cayman Unterborn, and Michael Way for their incredible support of my work.

To my local colleagues in Riverside, especially Maryjo Brounce, Timothy Lyons, and Eddie Schwieterman, thank you for all of the fascinating lessons you have taught me about the evolution of Earth's atmosphere and interior through time. I consider myself extremely fortunate to be part of such a diverse and collegial departmental family.

Thanks to my colleagues within the Nexus for Exoplanet System Science (NExSS) network for many stimulating discussions. In particular, I would like to thank Daniel Apai, Rory Barnes, Shawn Domagal-Goldman, Vikki Meadows, and Mary Voytek for their leadership in pushing forward various habitability aspects of the NExSS community.

I am grateful to the IOP publishing team who have provided wonderful support in making this book happen. My sincere gratitude and apologies to Leigh Jenkins for tolerating my frequent stating that "I just need a few more months." You have the patience of a saint!

Last but definitely not least, I want to express my sincere appreciation to my current research group members: Kimberly Bott, Alma Ceja, Paul Dalba, Tara Fetherolf, Michelle Hill, Zhexing Li, Colby Ostberg, Emilie Simpson, and Pam Vervoort. They have had to listen to me complain about writing this book for far too long. I'm happy to say that your ordeal is now at an end!

Author Biography

Stephen R. Kane

Stephen R. Kane is a Professor of Planetary Astrophysics at the University of California, Riverside who specializes in exoplanetary science and planetary habitability. He grew up in outback Australia where his view of the night sky and fascination with solar system exploration motivated his eventual career path. He received his Bachelor of Science (Honours) from Macquarie University in Sydney and his Doctorate from the University of Tasmania. His work covers a broad range of topics related to planetary astrophysics and he has discovered and co-discovered hundreds of planets orbiting other stars. He is a leading expert on the topic of planetary habitability, the habitable zone of planetary systems, and the study of why Venus and Earth underwent divergent evolutions. He is a prominent scientific leader for several NASA missions designed to search for life in the universe. He has published hundreds of peer-reviewed scientific papers as well as several books on the topic of exoplanets and habitability. He is also a prolific advocate of interdisciplinarity science through the combination of biology, climate science, geophysics, planetary science, and stellar astrophysics. When not obsessing over his science interests, he enjoys playing board games, model building and painting, writing fiction about alternate history, trying to conceptualize the perfect maze algorithm, the puzzles of Ernő Rubik, and a good Pinot noir while speculating about the nature of reality.

Planetary Habitability

Stephen R Kane

Chapter 1

Introduction

1.1 Why Does Habitability Matter?

Perhaps the most prevalent scientific question that has passed down through the ages is the speculation of life beyond Earth and how our ecosystem fits within that broader landscape of potential biodiversity. Although confined to the domain of philosophers for the majority of human history, technology has caught up to many of the proposed ideas for exolife detection in recent years, and the field of astrobiology is now a flourishing endeavor (Catling et al. 2018; Fujii et al. 2018; Kiang et al. 2018; Meadows et al. 2018; Schwieterman et al. 2018; Walker et al. 2018). The progress within astrobiology has been enabled by more than astronomy and biology, but also encompasses the disciplines of chemistry, climate science, geology, and planetary science, to name a few. The connection between these disciplines has produced a system science approach that recognizes the intricate dependencies between the various facets of planetary evolution (Kasting & Catling 2003; Lammer et al. 2013; Seager 2013; Dehant et al. 2019). Thus, after many centuries of speculation, we are entering an era in which the search for life in the universe may be quantified.

Even so, there exists a significant challenge of avoiding an anthropic view of how planetary habitability and, more broadly, astrobiology should be addressed. Much has been written on the subject of Earth as a prototype for exoplanet observations and associated detectable signatures (Bétrémieux & Kaltenegger 2013; Arney et al. 2017; Fan et al. 2019). Shown in Figure 1.1 are the predicted reflective properties of an Earth analog as a function of wavelength for a variety of surface properties (Fujii et al. 2010, 2011). Analyses such as these are being used to guide predictions for exoplanets, and thus instrument and mission design for the next generation of direct imaging and spectroscopy capabilities. The hope is that this research and development trajectory is the optimal pathway toward the detection of signatures of biological activity occurring on an exoplanet (so-called "biosignatures"). However, at the present time, Earth is the only location known to have a sustained biological

doi:10.1088/2514-3433/ac2aa1ch1

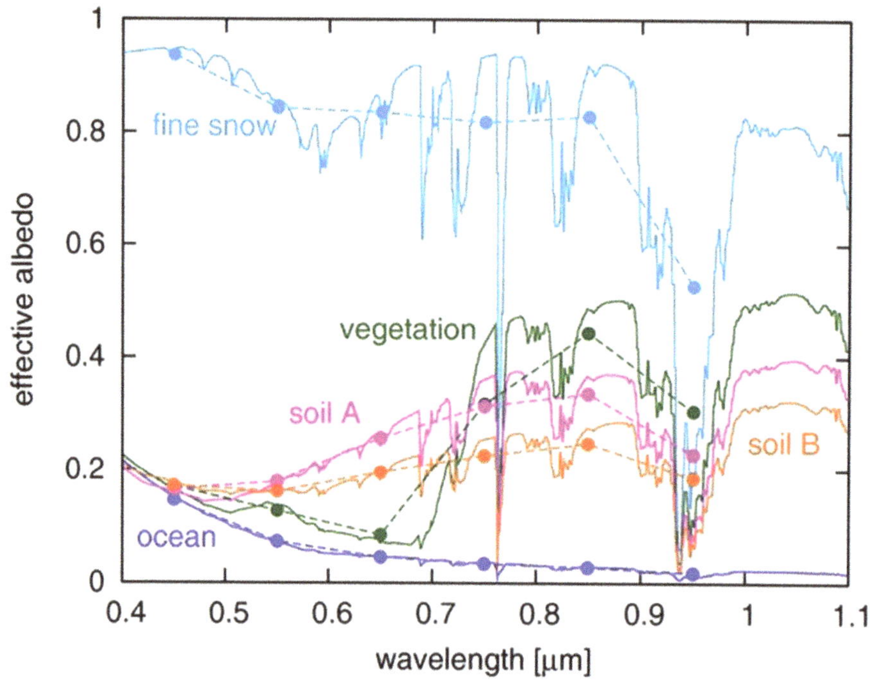

Figure 1.1. Wavelength-dependent surface reflectivity (albedos) of ocean (blue), soil (magenta/orange), vegetation (green), snow (cyan), assuming an Earth-like atmosphere and neglecting clouds. Credit: Reproduced from Fujii et al. (2011). © 2011. The American Astronomical Society. All rights reserved.

presence, although the possibility of life has been considered for Mars (Farmer & Des Marais 1999; Krasnopolsky et al. 2004), Europa (Kargel et al. 2000; Russell et al. 2017), Enceladus (Parkinson et al. 2007; McKay et al. 2008) and even Venus (Cockell 1999; Greaves et al. 2021). The biological data available in the solar system may therefore be considered limited within the universal context of the aforementioned biodiversity.

Thus, there are numerous reasons planetary habitability is a topic of great importance. First, understanding the factors that affect temperate planetary surface conditions is critical for studying the evolution of the solar system terrestrial bodies, most particularly Venus, Earth, and Mars. Second, a full appreciation of these factors will enable predictive capabilities with regard to the habitability of planets in general, including those outside of our solar system. Third, predicting exoplanet habitability will form the basis of target selection for current and future telescope resources in the search for possible biosignatures. Last but not least, an intricate study of planetary habitability will allow us to become better stewards of our own planet, whose surface conditions and ecosystem are clearly vulnerable to the collective behavior of our species. Thus, the purpose of studying habitability can be encapsulated within the combined quest of searching for other life throughout the universe, and the preservation of the singular biological ecosystem that the universe is known to have produced.

1.2 Defining Habitability

Defining the precise meaning of planetary habitability is an extremely difficult topic and, at the present time, remains controversial. This is not just because of the vast number of factors that influence planetary habitability, but also because it is very difficult to determine where the boundaries of habitability lie. Defining such boundaries makes the claim that bodies outside of those boundaries are uninhabitable and it is almost impossible to state unequivocally that a particular environment is completely hostile to all forms of biological chemistry. Additionally, a suitable definition of habitability can depend on the science goals of the end user. For example, an astronomer may prefer a definition that allows such habitability to be reliably detected at a distance of at least several parsecs, whereas an ecologist may favor a definition that explores the macroscopic details of a local environment.

At the base level, one may consider habitability to be defined as "an environment that is capable of sustaining life." Though not inaccurate, this definition could be deemed sufficiently broad and vague so as to have limited practical application, particularly as definitions of life itself are not globally agreed upon (Cleland & Chyba 2002; Benner 2010). Therefore, it is worth investigating the concepts of detectable versus non-detectable habitability as pre-requisites for a suitable definition. To be clear, detectability assumes remote detection since it is highly unlikely that *in situ* data for an exoplanet will be available any time soon, and probably for many generations. As such, detection of a habitable environment will be limited to inferences made for a planet, most particularly the planetary atmosphere and attempts to model the surface conditions. It is also worth pointing out that the detection of biosignatures for a particular planet implies (by definition) habitable conditions, but the inference of planetary habitability does not imply the presence of a biological ecosystem for the planet, nor that biosignatures will be present or detectable (Tasker et al. 2017).

A central planetary property that has dominated the discourse surrounding planetary habitability thus far is the presence of surface liquid water. For example, one of the defining features of Earth through time is the sustained presence of surface liquid water throughout most of Earth history, and the critical role this has played in the development of life on Earth (Brack 1993). Sustaining surface water in a liquid state over ~4 billion years is no small task given the relatively small range of temperatures required (see Figure 1.2). Surface liquid water has therefore gained a central role as a necessary but insufficient requirement for planetary habitability, and is also a core component in defining the habitable zone (HZ) of a star (Kasting et al. 1993; Kane & Gelino 2012; Kopparapu et al. 2013, 2014), described in Chapter 6. When considering water as a prerequisite for general planetary habitability prospects, there lies an intrinsic assumption regarding the universal dependence of biological development on water. This is not an unreasonable assumption given the prevalence of water (Simpson 2017; Carr & Najita 2011) and its properties of high heat capacity, stability in the presence of oxygen, and a neutral solvent for biochemistry, but it is an assumption nonetheless. However, given its role in the development of life of Earth, requiring surface liquid water provides an interesting framework from which we can study the effects of planetary properties and its environment on the changing surface

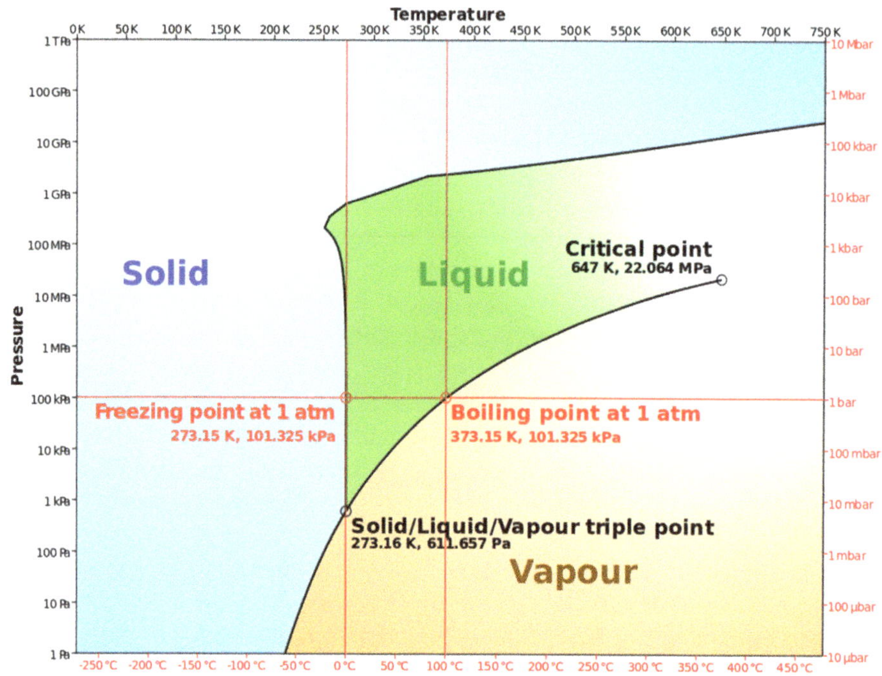

Figure 1.2. Pressure/temperature phase change diagram for water, demonstrating the limitations within which water can exist in a liquid state. For pressures lower than 1 atmosphere, the range of possible temperatures rapidly diminishes for surface liquid water. Credit: Wikimedia: Cmglee: CC BY-SA 3.0.

conditions of the planet. The utility of this framework is to enable a quantification of factors that would cause the surface temperature to fall outside the range of liquid water, which places the focus on physical processes and their effects relative to each other. We will explore these factors that influence surface conditions, and hence planetary habitability, in further detail in Chapter 2.

1.3 The Terrestrial Planet Demographic

Discussions regarding habitable planets outside of our solar system and their possible discovery have persisted for many years (Dole 1964). However, it is only over the past few decades that exoplanet discoveries have enabled the gradual transformation of these discussions into a data-driven realm. At the present time, the number of known exoplanets is rapidly approaching 5000, and these planetary systems have been shown to exhibit a vast diversity of architectures (Ford 2014; Winn & Fabrycky 2015; Hatzes 2016; He et al. 2019). Many of the observed orbital configurations are vastly different from those in the solar system. Shown in Figure 1.3 is a plot of the planetary mass and orbital period for many of the known exoplanets. The data shown were extracted from the NASA Exoplanet Archive[1]

[1] https://exoplanetarchive.ipac.caltech.edu/

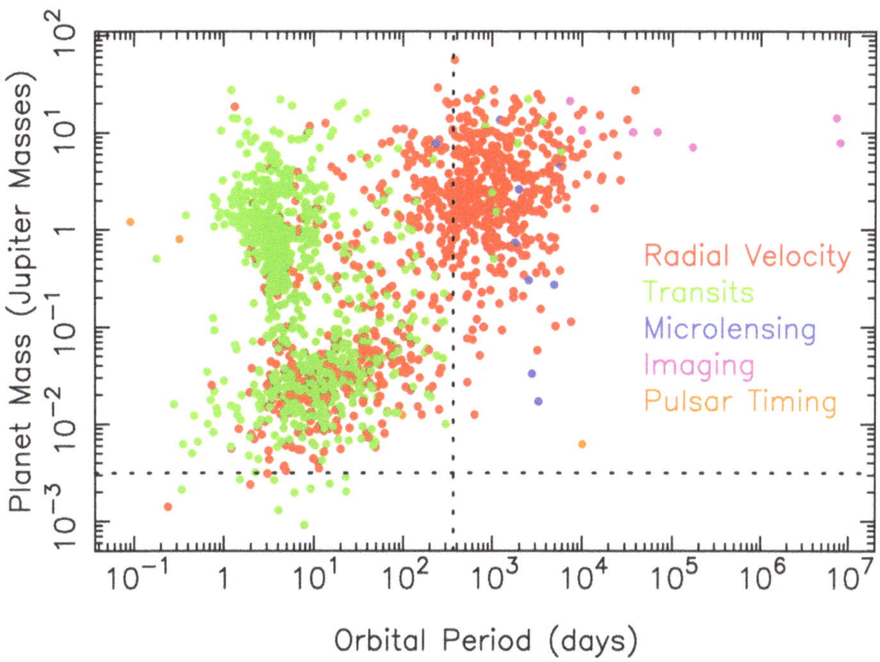

Figure 1.3. Planetary mass and orbital period for the known exoplanets, as of 2021 June 28, shown color-coded by the method of discovery for the individual planets. The dotted lines indicate the mass and orbital period of Earth. The relative lack of planets toward the bottom right of the plot (low-mass planets at long orbital periods) is due to the current sensitivity limits for each method, rather than an actual dearth of such planets. The data were extracted from the NASA Exoplanet Archive (Akeson et al. 2013).

(Akeson et al. 2013), where the data are current as of 2021 June 28. The vast majority of exoplanets have been detected using the radial velocity (sensitive to planet mass) and transit (sensitive to planet radius) methods, shown in red and green, respectively. Exoplanet detection methods, such as the radial velocity technique, are now able to probe into the regime of long-period Jupiter analogs (Wittenmyer et al. 2020). Discoveries have also included planets that have no solar system analogs, such as super-Earths (Valencia et al. 2007; Léger et al. 2009; Howard et al. 2010; Bonfils et al. 2013), mini-Neptunes (Barnes et al. 2009; Lopez & Fortney 2014; Nielsen et al. 2020), and hot Jupiters (Mayor & Queloz 1995; Fortney et al. 2008; Wright et al. 2012). Though most of the planet discoveries have occurred using methods that are biased toward short-period orbits, it is increasingly clear that the demographic of system architectures is both varied and complex.

The primary watershed moment for the discovery of many terrestrial planets arrived with the success of the NASA Kepler mission (Borucki et al. 2010). A primary goal of Kepler was to determine how common terrestrial planets are within the HZ of stars similar to the Sun; information of great interest for the field of astrobiology, as this is where a potential Earth analog with surface liquid water may reside. The Kepler telescope was launched in 2009, and operated nominally for four

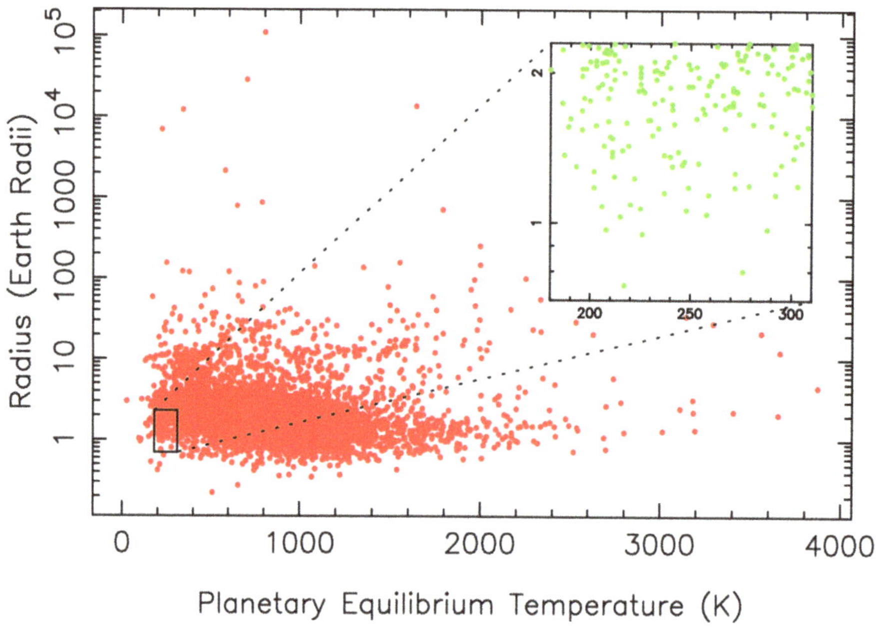

Figure 1.4. Planetary radius and calculated equilibrium temperature for more than 4,600 exoplanet candidates discovered via the Kepler mission. The zoomed inset box highlights those planets that may be both terrestrial and with equilibrium temperatures similar to Earth. The data were extracted from the NASA Exoplanet Archive (Akeson et al. 2013).

years, during which it stared at a large patch of sky and obtained precision photometry for ~170,000 stars. The self-consistent nature of the data allowed for detailed statistical analyses of exoplanet occurrence rates, and showed that terrestrial planets are far more common than their larger giant planet siblings (Winn & Fabrycky 2015; Borucki 2016). Figure 1.4 shows the planetary radius measurements and calculated equilibrium temperatures for more than 4600 Kepler exoplanet candidates that were detected during the primary mission. The inset box provides a zoomed in view for those candidates that may be terrestrial in nature and whose equilibrium temperatures are similar to Earth. It should be noted that the equilibrium temperature ignores many of the planetary properties that influence surface temperatures, most particularly the atmosphere (see Chapter 2). Even so, a catalog of the Kepler exoplanet inventory demonstrated that many of the Kepler candidates do indeed lie within the HZ of their host stars, motivating the need for further study of these planets (Kane et al. 2016). Undoubtedly, the main legacy of the Kepler mission is the extraordinary expansion of exoplanetary science into the realm of terrestrial planets, demonstrating that such planets are indeed extremely common. This realization paved the motivation for a deeper understanding of planetary processes, including their interiors, atmospheres, and climates.

The rapid push of exoplanetary science into the terrestrial planet regime has also further increased interest in defining the parameters of potentially habitable environments.

Exoplanet discoveries are continuing to occur via observations from the Transiting Exoplanet Survey Satellite (TESS; Ricker et al. 2015). TESS is surveying most of the bright stars in the entire sky, enabling mass measurements and atmospheric studies that are untenable for the majority of Kepler exoplanets. With so many terrestrial exoplanets now available for detailed studies, it is expected that the years ahead will provide vast amounts of terrestrial exoplanet data that can be compared with the rocky bodies of our solar system, including Titan (Horner et al. 2020; Kane et al. 2021). Terrestrial exoplanets will form the basis for a large-scale effort toward measurements of planetary atmospheric characteristics (Greene et al. 2016; Batalha et al. 2018; Kempton et al. 2018; Lustig-Yaeger et al. 2019) and inferring surface conditions (Wolf et al. 2017). The knowledge that rocky planets are extremely common, and the prospects for characterizing their atmospheres, has enabled an unprecedented quantification of possible surface environments (Shields 2019). Thus, it is the combination of the *in situ* data from our local (solar system) environment with the statistical power of terrestrial exoplanets that the true nature of planetary habitability will eventually be realized.

References

Akeson, R. L., Chen, X., Ciardi, D., et al. 2013, PASP, 125, 989

Arney, G. N., Meadows, V. S., Domagal-Goldman, S. D., et al. 2017, ApJ, 836, 49

Barnes, R., Jackson, B., Raymond, S. N., West, A. A., & Greenberg, R. 2009, ApJ, 695, 1006

Batalha, N. E., Lewis, N. K., Line, M. R., Valenti, J., & Stevenson, K. 2018, ApJ, 856, L34

Benner, S. A. 2010, AsBio, 10, 1021

Bétrémieux, Y., & Kaltenegger, L. 2013, ApJ, 772, L31

Bonfils, X., Delfosse, X., Udry, S., et al. 2013, A&A, 549, A109

Borucki, W. J. 2016, RPPh, 79, 036901

Borucki, W. J., Koch, D., Basri, G., et al. 2010, Sci, 327, 977

Brack, A. 1993, OLEB, 23, 3

Carr, J. S., & Najita, J. R. 2011, ApJ, 733, 102

Catling, D. C., Krissansen-Totton, J., Kiang, N. Y., et al. 2018, AsBio, 18, 709

Cleland, C. E., & Chyba, C. F. 2002, OLEB, 32, 387

Cockell, C. S. 1999, P&SS, 47, 1487

Dehant, V., Debaille, V., Dobos, V., et al. 2019, SSRv, 215, 42

Dole, S. H. 1964, Habitable Planets For Man (Waltham, MA: Blaisdell Publishing Company)

Fan, S., Li, C., Li, J.-Z., et al. 2019, ApJ, 882, L1

Farmer, J. D., & Des Marais, D. J. 1999, JGR, 104, 26977

Ford, E. B. 2014, PNAS, 111, 12616

Fortney, J. J., Lodders, K., Marley, M. S., & R S Freedman, 2008, ApJ, 678, 1419

Fujii, Y., Angerhausen, D., Deitrick, R., et al. 2018, AsBio, 18, 739

Fujii, Y., Kawahara, H., Suto, Y., et al. 2011, ApJ, 738, 184

Fujii, Y., Kawahara, H., Suto, Y., et al. 2010, ApJ, 715, 866

Greaves, J. S., Richards, A. M. S., Bains, W., et al. 2021, NatAs, 5, 655

Greene, T. P., Line, M. R., Montero, C., et al. 2016, ApJ, 817, 17

Hatzes, A. P. 2016, SSRv, 205, 267

He, M. Y., Ford, E. B., & Ragozzine, D. 2019, MNRAS, 490, 4575

Horner, J., Kane, S. R., Marshall, J. P., et al. 2020, PASP, 132, 102001
Howard, A. W., Marcy, G. W., Asher, J., Johnson, J. A., et al. 2010, Sci., 330, 653
Kane, S. R., & Gelino, D. M. 2012, PASP, 124, 323
Kane, S. R., Hill, M. L., Kasting, J. F., et al. 2016, ApJ, 830, 1
Kane, S. R., Arney, G. N., Byrne, P. K., et al. 2021, JGRE, 126, e06643
Kargel, J. S., Kaye, J. Z., Head, J. W., et al. 2000, Icar, 148, 226
Kasting, J. F., & Catling, D. 2003, ARA&A, 41, 429
Kasting, J. F., Whitmire, D. P., & Reynolds, R. T. 1993, Icar, 101, 108
Kempton, E. M. R., Bean, J. L., Louie, D. R., et al. 2018, PASP, 130, 114401
Kiang, N. Y., Domagal-Goldman, S., Parenteau, M. N., et al. 2018, AsBio, 18, 619
Kopparapu, R. K., Ramirez, R., Kasting, J. F., et al. 2013, ApJ, 765, 131
Kopparapu, R. K., Ramirez, R. M., SchottelKotte, J., et al. 2014, ApJ, 787, L29
Krasnopolsky, V. A., Maillard, J. P., & Owen, T. C. 2004, Icar, 172, 537
Lammer, H., Blanc, M., Benz, W., et al. 2013, AsBio, 13, 793
Léger, A., Rouan, D., Schneider, J., et al. 2013, A&A, 506, 287
Lopez, E. D., & Fortney, J. J. 2014, ApJ, 792, 1
Lustig-Yaeger, J., Meadows, V. S., & Lincowski, A. P. 2019, AJ, 158, 27
Mayor, M., & Queloz, D. 1995, Natur, 378, 355
McKay, C. P., Porco, C. C., Altheide, T., Davis, W. L., & Kral, T. A. 2008, AsBio, 8, 909
Meadows, V. S., Reinhard, C. T., Arney, G. N., et al. 2018, AsBio, 18, 630
Nielsen, L. D., Gandolfi, D., Armstrong, D. J., et al. 2020, MNRAS, 492, 5399
Parkinson, C. D., Liang, M. C., Hartman, H., et al. 2007, A&A, 463, 353
Ricker, G. R., Winn, J. N., Vanderspek, R., et al. 2007, JATIS, 1, 014003
Russell, M. J., Murray, A. E., & Hand, K. P. 2017, AsBio, 17, 1265
Schwieterman, E. W., Kiang, N. Y., Parenteau, M. N., et al. 2018, AsBio, 18, 663
Seager, S. 2013, Sci., 340, 577
Shields, A. L. 2019, ApJS, 243, 30
Simpson, F. 2017, MNRAS, 468, 2803
Tasker, E., Tan, J., Heng, K., et al. 2017, NatAs, 1, 0042
Valencia, D., Sasselov, D. D., & O'Connell, R. J. 2007, ApJ, 665, 1413
Walker, S. I., Bains, W., Cronin, L., et al. 2018, AsBio, 18, 779
Winn, J. N., & Fabrycky, D. C. 2015, ARA&A, 53, 409
Wittenmyer, R. A., Wang, S., Horner, J., et al. 2020, MNRAS, 492, 377
Wolf, E. T., Shields, A. L., Kopparapu, R. K., Haqq-Misra, J., & Toon, O. B. 2017, ApJ, 837, 107
Wright, J. T., Marcy, G. W., Howard, A. W., et al. 2012, ApJ, 753, 160

Planetary Habitability

Stephen R Kane

Chapter 2

Habitability Factors

2.1 An Intricate Puzzle

A fundamental challenge in modeling the atmospheres, interiors, and surfaces of planets lies in understanding the overall energy balance that determines the temperature structure throughout the planet. The sources of energy that contribute to planetary temperatures are many and varied, and are both internal and external to the planet. For the topic of planetary habitability, we are primarily concerned with the sources of energy that determine and control the surface temperature of a terrestrial planet. As described in Section 1.2, the temperature range that allows for surface liquid water is frequently used as the requirement for habitability. This liquid water temperature range need not be global (as it isn't for Earth) and one can consider the fraction of a planetary surface for which the liquid water temperature range is a valid assumption (Jansen et al. 2019).

The central theme of surface liquid water provides a quantifiable framework from which all the factors that control the surface temperature of a planet may be considered. A simplistic approach is to treat the planet as a blackbody and calculate the equilibrium temperature of a planet, T_{eq}. This is given by the following expression:

$$T_{eq} = \left(\frac{L_\star(1 - A_B)}{16\pi\sigma a^2} \right)^{\frac{1}{4}} \tag{2.1}$$

where $L_\star$ is the stellar luminosity, A_B is the spherical (Bond) albedo of the planet, and a is the star–planet separation. The luminosity of the star may be calculated using the Stefan–Boltzmann law:

$$L_\star = 4\pi R_\star^2 \sigma T_{\rm eff}^4 \tag{2.2}$$

where $R_\star$ is the stellar radius, $T_{\rm eff}$ is the stellar effective temperature, and σ is the Stefan–Boltzmann constant. Equation (2.1) applies to a planet that reradiates over 4π steradians, and thus assumes that the atmosphere is 100% efficient at redistributing energy received at the dayside to the nightside. However, although the equilibrium

doi:10.1088/2514-3433/ac2aa1ch2

temperature includes the dependency on the incident radiation and the albedo, it fails to include many other factors (Del Genio et al. 2019). For that reason, Equation (2.1) is sometimes used as a first order estimate of planetary temperature, but rarely considered an accurate indicator of planetary habitability. Figure 2.1 provides a significantly more complete framework for habitability that divides the factors controlling surface liquid water into the three broad categories of planetary properties, stellar effects, and the planetary system (Meadows & Barnes 2018). The framework does not claim to be all encompassing, but it does include many subtle aspects of planetary habitability that are frequently neglected, many of which are dependent on each other in equally subtle fashion. For example, the orbital evolution of a system influences tides, the amplitude of which is, in turn, dependent upon the internal structure and composition of the planet. Indeed, Figure 2.1 is frequently referred to as the "planets are hard" diagram by the creators, Victoria Meadows and Rory Barnes, to highlight the complexity of attempting to gain a full understanding of the requirements for surface liquid water. It is worth noting that the factors shown in Figure 2.1 include both potentially positive and negative effects on habitability. For example, although the energy a planet receives from the star tends to dominate the energy budget, stellar activity can have a detrimental effect via harmful radiation and atmospheric erosion.

The intricate puzzle of understanding planetary habitability requires more than appreciating the full range of factors that influence surface temperature. To

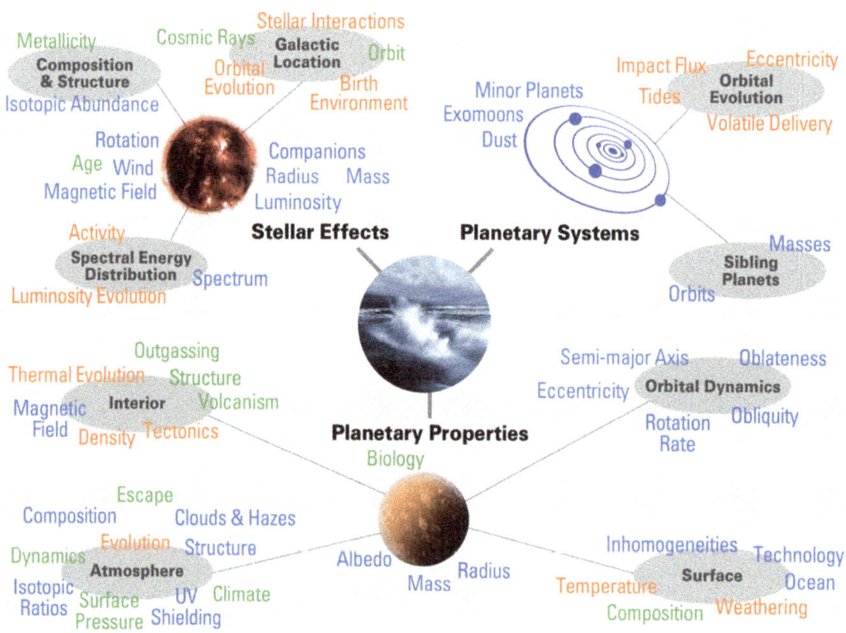

Figure 2.1. A summary of factors that control planetary surface temperature, and thus the presence of liquid water. These are divided into three broad categories of planetary properties, stellar effects, and the planetary system. Credit: Reprinted by permission from Springer Nature: Meadows & Barnes (2018), © Springer International Publishing AG, part of Springer Nature 2018.

Table 2.1. Summary of habitability factors across three broad categories. Factors shown in bold are largely accessible at the present time.

Category	Habitability factors
Stellar astrophysics	**Mass**, **Radius**, **Effective temperature**, **Luminosity**, **Composition**, **Rotation rate**, **Activity**, **Age**, **Multiplicity**
Planetary properties	**Mass**, **Radius**, **Bulk density**, Albedo, Atmospheric composition, Atmospheric dynamics, Pressure–temperature profile, **Age**
The planetary system	**Multiplicity**, **Semimajor axis**, **Eccentricity**, **Argument of periastron**, **Orbital inclination**, Rotation rate, Spin-axis obliquity, Exomoons, **Orbital evolution and stability**

complete the picture of potential habitability, we must also recognize which of these factors and processes dominate the controls over surface temperature. Furthermore, it is a sobering fact that many of these habitability factors will be extremely difficult to measure, and some will only be estimated from models based on solar system data, at least for the foreseeable future. Thus, the challenge of planetary habitability falls to answering three primary questions:

1. **What are all of the factors and processes that influence the long-term temperature environment of a planetary surface?**
2. **What is the relative weighting of these factors when assessing their influence on the surface environment?**
3. **How can these factors be measured or otherwise reliably estimated from models?**

For the remainder of this chapter, we briefly describe processes and questions related to three categories of habitability factors, similar to those shown in Figure 2.1. However, note that our discussion of orbit and spin dynamics for an individual planet will be incorporated into the over-arching theme of the orbital properties of a planetary system. A summary of some of the factors affecting habitability is shown in Table 2.1. Those factors shown in bold are those that are largely accessible via current measurements.

2.2 Stellar Astrophysics

As described in Section 2.1, a fundamental aspect of assessing the surface conditions of a planet, and thus the potential for long-term habitability, is establishing the energy balance that exists at the surface. By far, it is the influence of the host star that usually dominates the energy input into the overall energy balance. That is not to say that the presence of a nearby fusion reactor necessarily overwhelms a planet with positive contributions towards its habitability. The energy received at the top of the planetary atmosphere from the star often included wavelengths that may be considered harmful to many forms of biology. Additionally, some of those wavelengths will be of high enough energy (such as extreme ultraviolet, or XUV) to erode the atmosphere of the planet, particularly for active stars whose flares can have a devastating effect on planetary atmospheres (Tilley et al. 2019). Stars more massive than the Sun may also have relatively short life spans that reduce the relevance of possible habitability for harbored planets, and whose transitions through stellar evolution will likely destroy planets in close proximity. Thus, like many aspects of habitability, there may be considered stellar types that optimize the habitability prospects for an orbiting planet.

Many of the important stellar parameters required for studying planetary habitability are shown in the top row of Table 2.1. As can be seen from the highlighted parameters, all of those listed are accessible via current observations. For single stars, mass and radii are determined from stellar isochrone models (Dotter et al. 2008; Baraffe et al. 2015; Choi et al. 2016; Dotter 2016), although stellar radii may be measured directly via interferometry for nearby stars (van Belle & von Braun 2009; Boyajian et al. 2012). Although stellar effective temperatures are generally readily available via photometric colors and spectroscopy, detailed compositions require high-resolution spectra with high signal-to-noise (Hinkel et al. 2016). Likewise, stellar rotation rates are often accessible via precision photometry (Wright et al. 2011; McQuillan et al. 2014), although activity measurements usually require a more nuanced approach that connects with stellar spectra (Noyes et al. 1984; Saar & Donahue 1997; Isaacson & Fischer 2010). On the other hand, stellar ages (and thus planetary ages) are notoriously difficult for which to obtain reliable estimates (Kawaler 1988), and new techniques are being developed to improve age calculations (Angus et al. 2019). One of the most important stellar parameters for studying planetary habitability is the luminosity, since that determines the flux received at the top of the planetary atmosphere. Fortunately, the recent improvements to stellardistances provided by the Gaia astrometry mission (Gaia Collaboration et al. 2018) have enabled unprecedented improvements in stellar luminosities for a range of spectral types.

Although the host stars in planetary systems are generally better characterized than the planets themselves, numerous questions remain regarding stars and planetary habitability. Examples of these questions include:

- Can chemical abundances of the host star be used to constrain compositions of terrestrial planets?
- How does stellar multiplicity influence climate variations on exoplanets?
- What is the nature of stellar activity for young low-mass stars and can planetary atmospheres survive these periods?
- Can planets dynamically survive long-term stellar evolution, and what is the impact of such evolution on planetary surface conditions?

2.3 Planetary Properties

It could be argued that the intrinsic properties of a planet comprise the dominant components that govern the planetary surface conditions. For example, in much the same way that mass determines the bulk properties and evolution of a star, mass also plays a major role in the atmospheric and interior properties of a planet, as well as the evolution and processes that occur. Even so, it is difficult to know how processes, such as possible plate tectonics, scale with the mass of the planet (O'Neill & Lenardic 2007; Valencia et al. 2007) given the relative lack of available data for these processes within the solar system. Furthermore, the intricate dependencies of factors across all categories can result in significant model degeneracies in the correct interpretation of observable planetary properties.

A summary for some of the important planetary properties within the context of planetary habitability are shown in the middle row of Table 2.1. Those properties highlighted in bold are factors that are readily available at the present time, either through direct measurements or inferences from models. Note that not all of these properties are available for all planets, largely depending on aspects such as the exoplanet detection method used and the brightness of the host star. Planetary radii are predominantly measured for those planets whose orbital path happens to transit the disk of the host star along our line of sight; a very small fraction of the entire exoplanet population (Borucki & Summers 1984). The aforementioned planetary mass is generally measured through either the radial velocity technique or through interactions of planets that may result in transit timing variations (Agol et al. 2005; Fischer et al. 2016). Although masses have been measured for many planets more massive than Earth through these techniques, the relatively small perturbative signals produced by terrestrial planets makes their masses challenging to extract from the relevant data, and so are comparatively few at the present time.

It may also be stated that the ultimate goal of planetary habitability studies lies in the extraction of reliable information regarding the atmosphere, since it is the composition (particularly the absorption and emission properties of the species present) and pressure at the surface that determines the state of any water present. Moreover, the presence of water may be betrayed by signatures within the atmosphere itself, depending on the hydration and climate cycle of the atmosphere (Wolf et al. 2017). However, large-scale atmospheric detection for terrestrial planets will rely on future facilities, beginning with those accessible by the James Webb Space Telescope (JWST; Batalha et al. 2018; Kempton et al. 2018). These atmospheric measurements will in turn depend upon reliable mass measurements, since these play a critical role in constraining the atmospheric scale height (Batalha et al. 2019).

Despite the many challenges in extracting planetary properties in the terrestrial planet regime, it is helpful to define the scientific questions we seek to address in the context of habitability. Examples of these include:

- How small/large can a planet be and sustain long-term habitable conditions?
- How does outgassing scale with initial volatile content and internal heat budget?
- How do surface recycling processes scale with planet mass and composition?
- What information is required to predict atmospheric pressure–temperature profiles that extend to the surface?

2.4 The Planetary System

In determining the total energy budget of a planet, the location and kinematics of the planet, including external influences, provide another major source of influencing factors. These not only affect the energy received by the planet and how it is distributed over the surface via rotation and obliquity, but also present a time dependent component to the incident flux. Such time dependency can originate from several factors, most commonly the ellipticity of the orbit that causes the star–planet separation to change during the orbit. This time component can result in significant variability of the incident flux which, when coupled with the atmospheric dynamics, may result in climate variations that can truncate habitability prospects for the planet. Furthermore, tidal effects due to the host star, harbored moons, or other planets in the system, can inject an additional energy source into the overall energy budget of the planet. The additional tidal energy can sustain long-lived habitable surface conditions, or may result in catastrophic climate collapse due to excessive energy release from the interior. The orbital architecture of the system can also be used to study the formation processes and potentially predict migration and volatile delivery scenarios.

Many of the major factors affecting habitability resulting from orbits and interactions between bodies are shown in the bottom row of Table 2.1. Determining the full architecture (multiplicity) of planetary systems, including long-period planets, is the ongoing goal of several long-term monitoring programs (Winn & Fabrycky 2015; Wittenmyer et al. 2020). The semimajor axis of a planet, crucial in calculating incident flux, is accessible via most exoplanet detection methods. Keplerian orbital parameters, such as eccentricity and argument of periastron, are usually extracted via radial velocity measurements (Halbwachs et al. 2005; Shen & Turner 2008) and occasionally the transit method (Ford et al. 2008; Kipping 2008; Kane et al. 2012; Van Eylen & Albrecht 2015). Keplerian orbits and planetary masses allow investigations into the dynamical interactions between planets (Smith & Lissauer 2009; Kane 2019) and possible tidal influences (Barnes et al. 2009). Though extremely important for climate modeling of terrestrial planets (Joshi 2003; Kane & Torres 2017), parameters related to the angular moment vector of the planet remain largely inaccessible at the present time (Cowan et al. 2013).

As we continue to diagnose the distribution of exoplanetary system architectures, the relevance of these distributions to habitability consequences become increasingly important. Examples of relevant questions include:

- What are the formation/dynamical implications of a giant planet (or planets) in the outer regions of a system?
- Can planets in extreme eccentric orbits retain surface habitability?
- How do terrestrial atmospheres respond to flux variations and over what time scales?
- Does the tidal influence of a substantial moon play a major role in habitability?

2.5 A Statistical Approach

The large number of known exoplanets has enabled detailed statistical studies, such as estimates for the occurrence rates of terrestrial planets that lie within the habitable zone (HZ; Dressing & Charbonneau 2013; Kopparapu 2013; Dressing & Charbonneau 2015; Bryson et al. 2021) and the dependence on stellar properties (Fischer & Valenti 2005; Buchhave et al. 2014). In addition to studying individual systems, the rapid increase in the number of known exoplanets also provides an opportunity to explore the distribution of parameters that are relevant to planetary

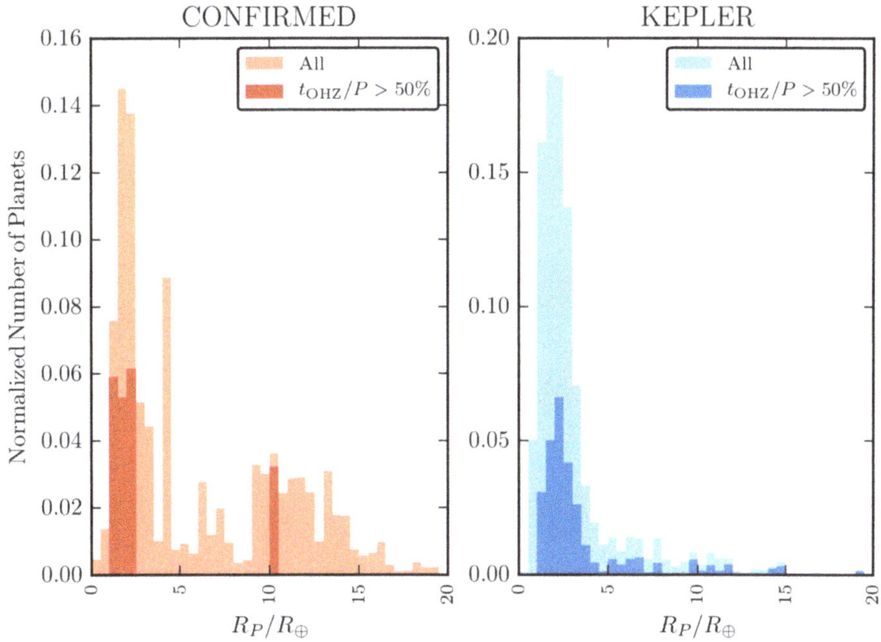

Figure 2.2. Distribution of transiting confirmed and Kepler candidate planets with respect to planetary radius. The data are color-coded according to the percentage of time each planet spends in its HZ. Credit: Reproduced from Adams & Kane (2016). © 2016. The American Astronomical Society. All rights reserved.

habitability. For example, Figure 2.2 shows histograms of planetary radii for confirmed transiting planets and Kepler candidates, color-coded by whether they reside within the HZ or not (Adams & Kane 2016). These histograms further demonstrate the relative prevalence of terrestrial planets compared with gas giant planets (Borucki 2016). Such studies must account for observational bias, both in the selection of stellar targets (Santos et al. 2000; Brown et al. 2011; Butler et al. 2017; Stassun et al. 2018, 2019) and biases inherent to the detection method (Zakamska et al. 2011; Winn & Fabrycky 2015). As exoplanet surveys continue toward completeness of their viable parameter space, such as longer period planets, the distributions of the full range of habitability-relevant parameters will become gradually clearer.

A major focus of current exoplanet surveys is leveraging the complimentary nature of the transit and radial velocity methods toward understanding the planetary mass–radius relationship (Seager et al. 2007; Swift et al. 2012; Weiss & Marcy 2014). The extraction of bulk density measurements are combined with interior models to gain insight into the possible compositions, interior structures, and potential surface conditions of the planet (Adams et al. 2008; Dorn et al. 2015; Unterborn & Panero 2019). Shown in Figure 2.3 is an example mass–radius diagram that contains data for exoplanets smaller than $4.5R_\oplus$ and solar system planets as gray solid circles (Li Zeng et al. 2019). The figure shows solid-line contours of constant density that include the effects of inflation for substantial H_2 atmospheres. The data are transposed to a radius histogram on the y-axis that exhibits a radius gap at

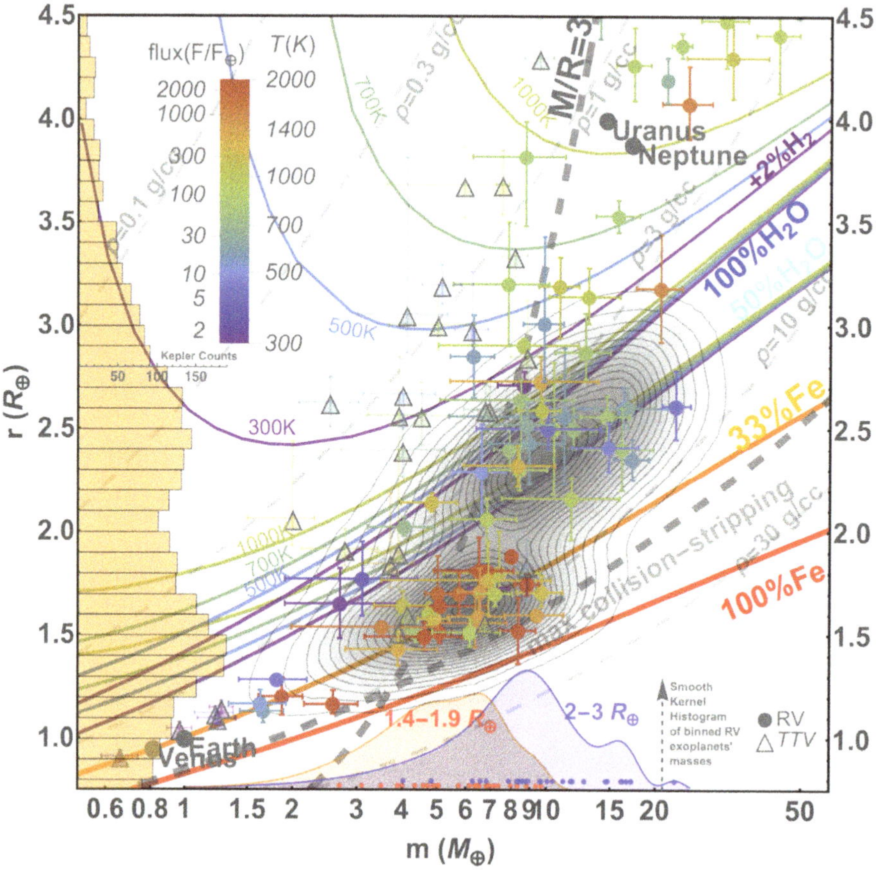

Figure 2.3. Mass–radius diagram for exoplanets smaller than $4.5R_\oplus$, showing a radius gap at $\sim 2R_\oplus$. The data points are color-coded by incident flux and solid-line contours of constant density are included. Credit: Reproduced from Zeng et al. (2019) with permission from Proceedings of the National Academy of Sciences of the United States of America.

$\sim 2R_\oplus$. The populations above and below this radius gap produce a similar bimodal mass distribution, shown on the x-axis. Understanding the nature of the observed radius gap in the mass–radius relationship provides a fundamental insight into planet formation, migration, and evaporation processes (Lopez & Fortney 2013; Owen & Wu 2013; Fulton et al. 2017).

Up until several decades ago, studies of planetary habitability were limited to terrestrial data available in the solar system, such as Venus, Earth, Mars, and Titan. However, the vast number of exoplanets now provides a statistical tool that may be applied to the key factors that influence surface conditions. The critical importance of this application of exoplanet statistics is made apparent when considering questions of how frequently favorable habitability factors manifest in planetary systems. Example questions that can be addressed via exoplanet statistics are:

- How do exoplanet architectures vary as a function of stellar mass?
- How common are giant planets in the HZ?

- What is the distribution of bulk densities for terrestrial planets and does that correlate with stellar compositional abundances?
- Do terrestrial planets orbiting M dwarf stars typically retain long-term atmospheres?

It is expected that the coming years will witness a considerable increase in the number of exoplanet atmosphere detection and characterization, enabled through new techniques and facilities. These data will be examined within the context of the overall system properties and potentially be used to either directly or indirectly infer habitable surface conditions. Each such inference will provide a valuable habitability data point that will be scrutinized in terms of the habitability factors described above. In parallel, significant advances are occurring in atmospheric laboratory work (Moran et al. 2020), heliophysics and space weather (Airapetian et al. 2020), and the connections between exoplanets and planetary science (Kane et al. 2021). However, it is through an exhaustive census of the properties of exoplanets, their host stars, and the planetary systems they occupy that the true impact of habitability factors will become known.

References

Adams, A. D., & Kane, S. R. 2016, AJ, 152, 4
Adams, E. R., Seager, S., & Elkins-Tanton, L. 2008, ApJ, 673, 1160
Agol, E., Steffen, J., Sari, R., & Clarkson, W. 2005, MNRAS, 359, 567
Airapetian, V. S., Barnes, R., Cohen, O., et al. 2020, IJAsB, 19, 136
Angus, R., Morton, T. D., Foreman-Mackey, D., et al. 2019, AJ, 158, 173
Baraffe, I., Homeier, D., Allard, F., & Chabrier, G. 2015, A&A, 577, A42
Barnes, R., Jackson, B., Greenberg, R., & Raymond, S. N. 2009, ApJ, 700, L30
Batalha, N. E., Lewis, N. K., Line, M. R., Valenti, J., & Stevenson, K. 2018, ApJ, 856, L34
Batalha, N. E., Lewis, T., Fortney, J. J., et al. 2019, ApJ, 885, L25
Borucki, W. J. 2016, RPPh, 79, 036901
Borucki, W. J., & Summers, A. L. 1984, Icar, 58, 121
Boyajian, T. S., McAlister, H. A., van Belle, G., et al. 2012, ApJ, 746, 101
Brown, T. M., Latham, D. W., Everett, M. E., & Esquerdo, G. A. 2011, AJ, 142, 112
Bryson, S., Kunimoto, M., Kopparapu, R. K., et al. 2021, AJ, 161, 36
Buchhave, L. A., Bizzarro, M., Latham, D. W., et al. 2014, Natur, 509, 593
Butler, R. P., Vogt, S. S., Laughlin, G., et al. 2017, AJ, 153, 208
Choi, J., Dotter, A., Conroy, C., et al. 2016, ApJ, 823, 102
Cowan, N. B., Fuentes, P. A., & Haggard, H. M. 2013, MNRAS, 434, 2465
Del Genio, A. D., Kiang, N. Y., Way, M. J., et al. 2019, ApJ, 884, 75
Dorn, C., Khan, A., Heng, K., et al. 2015, A&A, 577, A83
Dotter, A. 2016, ApJS, 222, 8
Dotter, A., Chaboyer, B., Jevremović, D., et al. 2008, ApJS, 178, 89
Dressing, C. D., & Charbonneau, D. 2013, ApJ, 767, 95
Dressing, C. D., & Charbonneau, D. 2015, ApJ, 807, 45
Fischer, D. A., Anglada-Escude, G., Arriagada, P., et al. 2016, PASP, 128, 066001
Fischer, D. A., & Valenti, J. 2005, ApJ, 622, 1102

Ford, E. B., Quinn, S. N., & Veras, D. 2008, ApJ, 678, 1407
Fulton, B. J., Petigura, E. A., Howard, A. W., et al. 2017, AJ, 154, 109
Gaia CollaborationBrown, A. G. A., Vallenari, A., et al. 2018, A&A, 616, A1
Halbwachs, J. L., Mayor, M., & Udry, S. 2005, A&A, 431, 1129
Hinkel, N. R., Young, P. A., Pagano, M. D., et al. 2016, ApJS, 226, 4
Isaacson, H., & Fischer, D. 2010, ApJ, 725, 875
Jansen, T., Scharf, C., Way, M., & Del Genio, A. 2019, ApJ, 875, 79
Joshi, M. 2003, AsBio, 3, 415
Kane, S. R., Ciardi, D. R., Gelino, D. M., & von Braun, K. 2012, MNRAS, 425, 757
Kane, S. R., & Torres, S. M. 2017, AJ, 154, 204
Kane, S. R. 2019, AJ, 158, 72
Kane, S. R., Arney, G. N., Byrne, P. K., et al. 2021, JGRE, 126, e06643
Kawaler, S. D. 1988, ApJ, 333, 236
Kempton, E. M. R., Bean, J. L., Louie, D. R., et al. 2018, PASP, 130, 114401
Kipping, D. M. 2008, MNRAS, 389, 1383
Kopparapu, R. K. 2013, ApJ, 767, L8
Lopez, E. D., & Fortney, J. J. 2013, ApJ, 776, 2
McQuillan, A., Mazeh, T., & Aigrain, S. 2014, ApJS, 211, 24
Meadows, V. S., & Barnes, R. K. 2018, Factors Affecting Exoplanet Habitability (Berlin: Springer), 57
Moran, S. E., Hörst, S. M., Vuitton, V., et al. 2020, PSJ, 1, 17
Noyes, R. W., Hartmann, L. W., Baliunas, S. L., Duncan, D. K., & Vaughan, A. H. 1984, ApJ, 279, 763
O'Neill, C., & Lenardic, A. 2007, GeoRL, 34, L19204
Owen, J. E., & Wu, Y. 2013, ApJ, 775, 105
Saar, S. H., & Donahue, R. A. 1997, ApJ, 485, 319
Santos, N. C., Mayor, M., Naef, D., et al. 2000, A&A, 361, 265
Seager, S., Kuchner, M., Hier-Majumder, C. A., & Militzer, B. 2007, ApJ, 669, 1279
Shen, Y., & Turner, E. L. 2008, ApJ, 685, 553
Smith, A. W., & Lissauer, J. J. 2009, Icar, 201, 381
Stassun, K. G., Oelkers, R. J., Paegert, M., et al. 2019, AJ, 158, 138
Stassun, K. G., Oelkers, R. J., Pepper, J., et al. 2018, AJ, 156, 102
Swift, D. C., Eggert, J. H., Hicks, D. G., et al. 2012, ApJ, 744, 59
Tilley, M. A., Segura, A., Meadows, V., Hawley, S., & Davenport, J. 2019, AsBio, 19, 64
Unterborn, C. T., & Panero, W. R. 2019, JGRE, 124, 1704
Valencia, D., O'Connell, R. J., & Sasselov, D. D. 2007, ApJ, 670, L45
van Belle, G. T., & von Braun, K. 2009, ApJ, 694, 1085
Van Eylen, V., & Albrecht, S. 2015, ApJ, 808, 126
Weiss, L. M., & Marcy, G. W. 2014, ApJ, 783, L6
Winn, J. N., & Fabrycky, D. C. 2015, ARA&A, 53, 409
Wittenmyer, R. A., Wang, S., Horner, J., et al. 2020, MNRAS, 492, 377
Wolf, E. T., Shields, A. L., Kopparapu, R. K., Haqq-Misra, J., & Toon, O. B. 2017, ApJ, 837, 107
Wright, N. J., Drake, J. J., Mamajek, E. E., & Henry, G. W. 2011, ApJ, 743, 48
Zakamska, N. L., Pan, M., & Ford, E. B. 2011, MNRAS, 410, 1895
Zeng, L., Jacobsen, S. B., Sasselov, D. D., et al. 2019, PNAS, 116, 9723

AAS | IOP Astronomy

Planetary Habitability

Stephen R Kane

Chapter 3

Nature's Fusion Reactors

3.1 Key Stellar Parameters

The influence of stars on planetary habitability forms approximately one third of the factors in the "planets are hard" diagram, shown in Figure 2.1. In terms of their contribution to the energy budget of a planetary surface, stars are overwhelmingly the dominant contributor throughout most of a planet's lifetime. As such, understanding the properties of stars, including their activity, evolution, and multiplicity, is a crucial piece of the overall habitability puzzle for a given planet.

Among the numerous attributes of stars, there are six intrinsic properties that are foundational and from which most other properties may be derived. These intrinsic properties are:

- **Mass**. The mass of the star, usually described in solar masses ($M_\odot = 1.989 \times 10^{30}$ kg).
- **Radius**. The radius of the star, usually described in solar radii ($R_\odot = 6.957 \times 10^8$ m).
- **Effective temperature**. The temperature of a blackbody that emits an equivalent amount of radiation. The effective temperature of the Sun is $T_{\rm eff} = 5772$ K.
- **Luminosity**. The total power output from the stellar surface. The luminosity of the Sun is $L_\odot = 3.828 \times 10^{26}$ W.
- **Composition**. The elemental abundances relative to hydrogen, often referred to as "metallicity" for the specific case of iron abundance [Fe/H].
- **Age**. The stellar age, usually expressed in Gyr, or Myr for particularly young stars.

Each of these properties have relevance to planetary habitability. In particular, the integrated flux received at the top of the planetary atmosphere scales linearly with the stellar luminosity, and the wavelength distribution of the incident flux depends on the effective temperature. The stellar radius is related to the luminosity and effective temperature via the Stefan–Boltzmann law (Equation (2.2)). Certain stellar

doi:10.1088/2514-3433/ac2aa1ch3

compositional abundances are assumed to be key ingredients in the bulk composition of terrestrial exoplanets (Dorn et al. 2015; Hinkel & Unterborn 2018). Likewise, the age of planets in the system are usually approximately the same age as the star. Finally, the stellar mass is the primary determinant for the radius, effective temperature, luminosity, and total main-sequence lifetime of the star. For main sequence stars, the mass relates to the luminosity via the following expression:

$$\frac{L_\star}{L_\odot} = \left(\frac{M_\star}{M_\odot}\right)^x \tag{3.1}$$

where the exponent, x, is often assigned a value of 3.5, but varies slightly as a function of mass (Kuiper 1938). The total main-sequence lifetime of a star, $\tau_{\rm MS}$, may be approximated as:

$$\tau_{\rm MS} = 10^{10}\left(\frac{M_\star}{M_\odot}\right)^{2.5} \tag{3.2}$$

where $\tau_{\rm MS}$ is measured in years (Sackmann et al. 1993). For a solar mass star, Equation (3.2) yields a main-sequence lifetime of 10 billion years.

The fundamental challenge in stellar astrophysics is measuring each of these properties, or deriving them from other observables. The flowchart shown in Figure 3.1 demonstrates example relationships between observables (parallax,

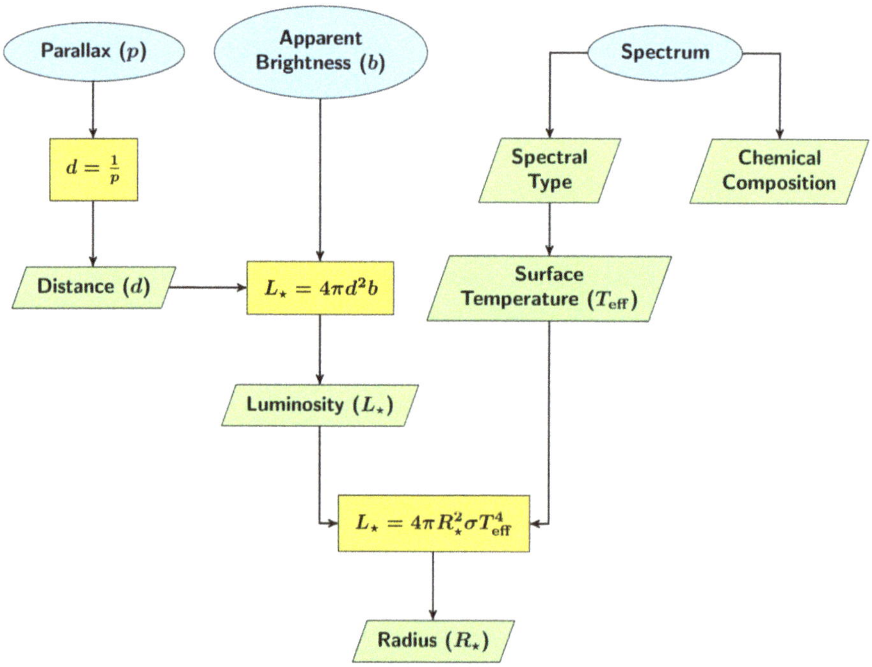

Figure 3.1. A stellar parameter flowchart that connects observables (shown in blue) to derived properties (shown in green) via known relationships (shown in yellow). Credit: Tara Fetherolf.

brightness, and spectra) and derived properties. The effective temperature can be extracted from photometric color information that assumes blackbody emission, although it is more reliably determined from spectra that account for absorption features in the stellar atmosphere. This is especially important for the relatively cool M dwarf stars, whose strong titanium oxide and sodium absorption bands can cause the spectral energy distribution to diverge significantly from that of an equivalent temperature blackbody (Bessell 1991; Boyajian et al. 2012; Muirhead et al. 2012). In principle, the stellar luminosity can be calculated using the Stefan–Boltzmann equation if a radius estimate is available via direct measurements or derived from isochrone models (see Section 2.2). However, the precision parallax measurements acquired by the Gaia mission have provided stellar distance estimates of unprecedented accuracy (Gaia Collaboration 2021). Assuming that the extinction along the line of sight is well understood, combining the distance with the apparent brightness of a star is a robust method to quantify the stellar luminosity, as shown in Figure 3.1.

A word on nomenclature regarding spectral types. Stars are generally classified under the Morgan–Keenan (MK) system (Johnson & Morgan 1953; Morgan & Keenan 1973) using the letters (in order of decreasing effective temperature) O, B, A, F, G, K, M. Each letter class is subdivided with numbers 0 through 9, also in order of decreasing effective temperature. For example, the Sun is classified as G2 star, Sirius (the brightest star in the night sky) is a A1 star, and Proxima Centauri (the nearest star to the Sun) is classified as M5.5. Searches for exoplanets tend to concentrate on lower mass, lower temperature stars in the spectral type range F–M. These stars have more spectral features than hotter stars, and these features enable spectroscopic exoplanet search techniques (Campbell & Walker 1979; Marcy & Butler 1992; Butler et al. 1996). Their lower stellar mass and radii also result in an orbiting planet having a greater, and therefore more measurable, effect. Furthermore, Equation (3.2) shows that a lower stellar mass means a significantly greater lifetime on the main sequence, consequently allowing more time for a potential biosphere to evolve. M dwarf stars, for example, have an estimated main-sequence lifetime that is larger than the current age of the universe, and thus we have never directly observed the fate of such a star when it exhausts its supply of hydrogen fuel. However, cooler stars can also have extended periods of high activity, which may be detrimental to planetary atmospheres and surface biology.

3.2 Stellar Activity and Atmospheric Erosion

Stellar activity is an exceptionally broad term that may be used to describe a variety of star-related phenomena. These phenomena primarily manifest in the stellar atmosphere and become apparent through changes in stellar brightness whose amplitudes peak in many different wavelengths, depending on the energy associated with the particular event behind the variability. Stellar activity is often caused by processes involving convection within the star and the interaction of the star's magnetic field with the stellar atmosphere (Noyes 1984; Baliunas & Vaughan 1985; Foukal et al. 2006; Wright et al. 2011). The extent of activity is also related to the age of the star, and therefore the age of any orbiting planets (Wilson 1963; Pace & Pasquini 2004; Kiraga & Stepien 2007). In terms of planetary habitability, there are two main effects of stellar activity: exoplanet detection and atmospheric erosion.

The detection of exoplanets generally relies on a comprehensive characterization of the stars being surveyed. For indirect methods, such as transits and radial velocities, the variability of stars can seriously impede the sensitivity of the data to exoplanet signatures, and may even be cast as false positive signatures (Santerne et al. 2013). For example, stellar pulsations can mimic planetary signatures in radial velocity data, and stellar jitter adds a non-periodic component to the noise floor of the data (Saar & Donahue 1997; Desort et al. 2007; Isaacson & Fischer 2010). Likewise, star spots, eclipsing binaries, and other photometric effects can impair the detection yield of transit surveys (Paulson et al. 2004; Newton et al. 2016) and may affect the derived planetary parameters (Czesla et al. 2009; Ciardi et al. 2015). As such, disentangling stellar activity and planetary signals is an important challenge, particularly for terrestrial planets whose signatures will often lie beneath the noise floor of the data (Boisse et al. 2011).

Research into the effect of stellar activity on planetary atmospheres has demonstrated that the activity can have a profound effect on the evolutionary pathway of planets (Lammer et al. 2003; Owen 2019; Rodríguez-Mozos & Moya 2019), particularly for those within the terrestrial regime (Dong et al. 2017a, 2018; Roettenbacher & Kane 2017; Sakuraba et al. 2019; Chen et al. 2021). This can be especially important for planets orbiting M dwarf stars, whose periods of high activity can persist substantially longer than for more massive stars (Delfosse et al. 1998), and they tend to remain relatively active through most of their main-sequence lifetimes (Newton et al. 2018). For example, the known exoplanet host star, Proxima Centauri, remains relatively active with frequent substantial flaring. Figure 3.2 shows Proxima Centauri observations from which a naked-eye-brightness superflare was detected, increasing the average brightness by a factor of 38, and the instantaneous brightness by a factor of 68 (Howard et al. 2018; MacGregor et al. 2018). The activity of Proxima Centauri may have substantially desiccated the atmosphere of any planets present in the system, reducing the likelihood of sustained habitability (Dong et al. 2017b; Meadows et al. 2018; Vida et al. 2019). An additional consequence of atmospheric erosion due to stellar activity can be the loss of water from the upper atmosphere through a process of water vapor photolysis and hydrogen/oxygen escape (Luger & Barnes 2015). Figure 3.3 provides a representation of water mass loss for an Earth-similar planet orbiting an M dwarf, as a function of orbital eccentricity (see Section 5.1) and incident flux (see Section 4.1; Palubski et al. 2020). For a circular orbit, the planet can lose the equivalent of Earth's entire surface water inventory in less than 1 Gyr. This is primarily due to the XUV flux received by the planet being a factor of 6 higher than that for a planet with an equivalent climate orbiting a G dwarf star. The result is consistent with others which have shown that, for stars less massive than the Sun, at least one Earth surface water

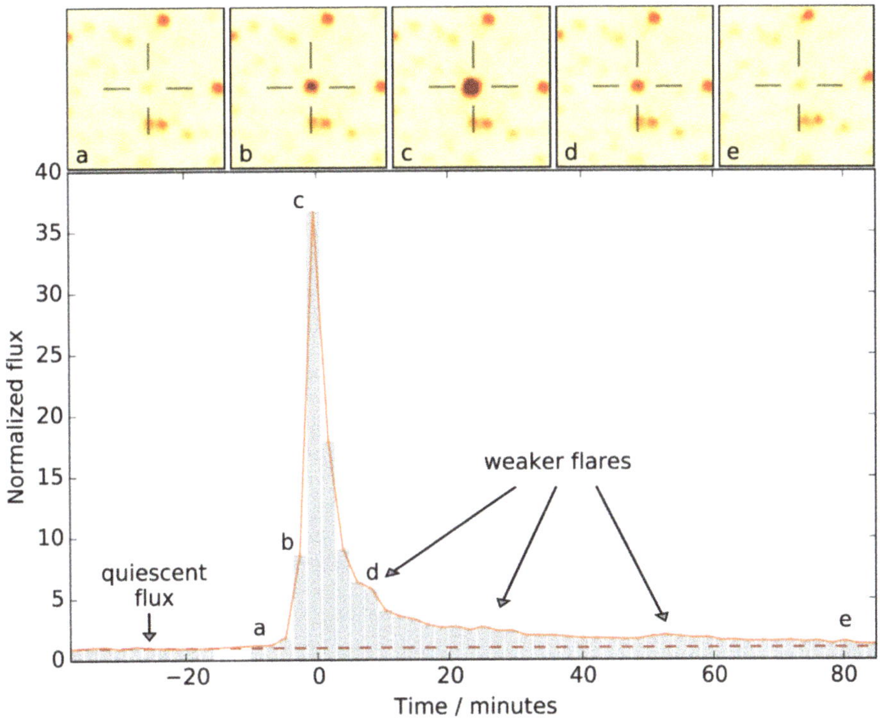

Figure 3.2. Brightness of Proxima Centauri as a function of time. Observations revealed several flares during a one hour period, including a naked-eye-brightness superflare. Bars show the integration time of each individual flux measurement, and the insets display the changing brightness of Proxima over the course of the observations. Credit: Reproduced from Howard et al. (2018). © 2018. The American Astronomical Society. All rights reserved.

inventory will have been lost after 5 Gyr for most locations within the HZ (see Chapter 6) due to atmospheric erosion (Luger & Barnes 2015).

3.3 Stellar Evolution

As stars continue to burn hydrogen during the main-sequence phase of their lives, several of the key parameters described in Section 3.1 undergo gradual but non-trivial changes. As the compositional ratio of hydrogen to helium in the stellar core changes, the energy output of the star naturally adjusts so that hydrostatic equilibrium is preserved. This results in an increase in luminosity, radius, and effective temperature of the star as it ages on the main sequence. Figure 3.4 shows this gradual change in parameters for the case of the Sun. The luminosity of the Sun as a function of time may be expressed as:

$$L_{\odot} = \frac{L_{\odot}(t_{\odot})}{1 + 0.4(1 - t/t_{\odot})} \tag{3.3}$$

where $t_{\odot} = 4.5$ Gyr is the present age of the Sun. Note the dramatic nature of the changes that occur in Figure 3.4 beyond ~10 Gyr, at which point the Sun will move

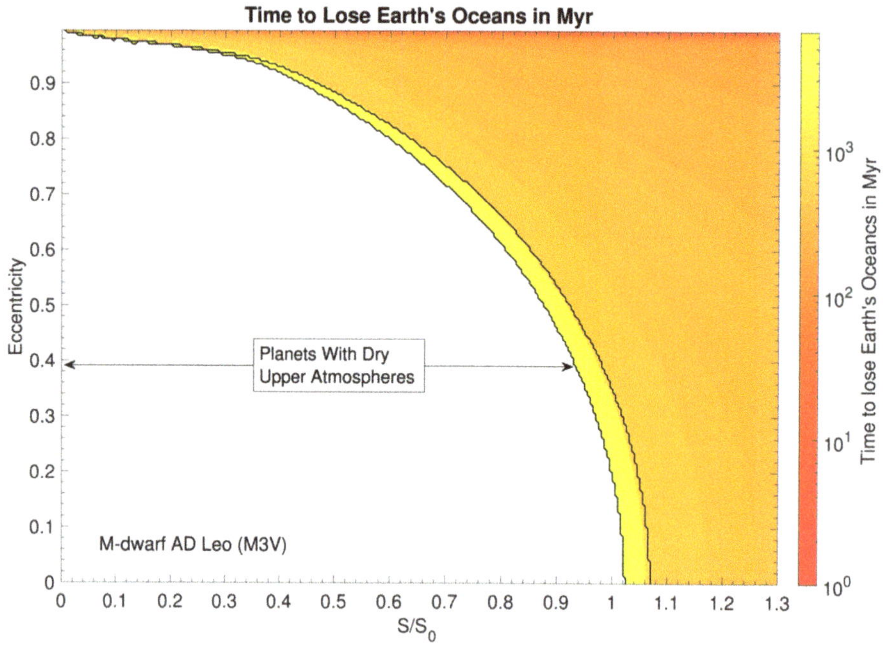

Figure 3.3. Water loss rate from an Earth-similar planet orbiting an M dwarf, as a function of orbital eccentricity and incident flux, in Earth units (S/S_0). Credit: Reproduced from Palubski et al. (2020).

off the main sequence into the red giant phase of its evolution. During the early history of the Earth, the Sun was ~25% less luminous than it is now. This "faint young Sun" has been proposed over many years as a paradox, since early calculations suggested it should have rendered the Earth in a frozen "snowball" state for the first 2 billion years (Feulner 2012). However, more recent calculations using 3D climate models indicate that the problem is resolved by substantially higher concentrations of CO_2 during the Archean, consistent with geological proxies (Charnay et al. 2020).

Of the parameters described in Section 3.1, the mass and the age are sufficient to provide robust estimates for most other properties of a star, since the mass is the primary determinant for the evolutionary pathway the star will follow. Evolutionary tracks for stars of various masses are shown in Figure 3.5, as a function of their luminosity and effective temperature (Hertzsprung–Russell diagram). The chemical abundance of stars also influence these tracks since, as described above, the energy output of the core required to achieve hydrostatic equilibrium is sensitive to composition. The tracks shown in Figure 3.5 all assume a metallicity similar to that of the Sun. Note also that the timescale of the pathways becomes dramatically shorter with increasing mass (see Equation (3.2)). The changes of stars on the evolutionary tracks results in a gradual change in the incident flux received at the top of a planetary atmosphere (Underwood et al. 2003; Ramirez & Kaltenegger 2014), described in more detail in Chapter 4.

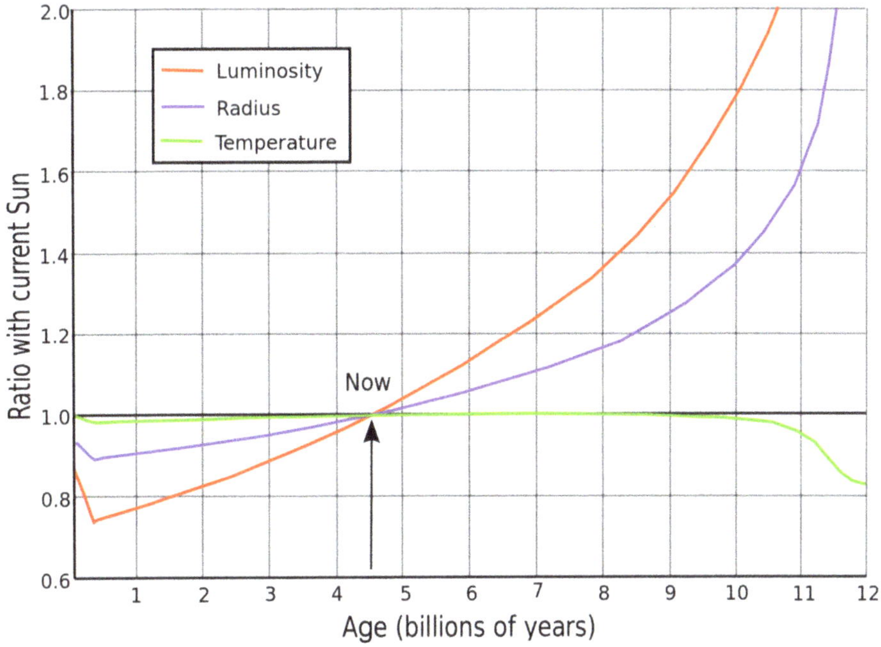

Figure 3.4. The evolution of the solar luminosity, radius, and effective temperature, expressed as a ratio of the current values. Credit: Wikimedia: R. J. Hall: CC BY-SA 3.0.

The known exoplanet hosts span a diverse range of spectral types and stages of stellar evolution. Figure 3.6 shows a Hertzsprung–Russell diagram for the known exoplanet hosts using data extracted from the Habitable Zone Gallery[1] (Kane & Gelino 2012), where the data are current as of 2020 December 28. The dashed crosshairs show the locations of the Sun, and the colors indicate the location of planets in the system relative to the habitable zone (HZ). Green points represent host stars that harbor a planet that spends more than 50% of the orbit within the HZ of the star. Note that the HZ planets represented here include giant planets as well as terrestrial. The abundance of red points (planets interior to the HZ) is reflective of the bias of present detection techniques towards planets close to their host stars (see Section 2.5). Several of the HZ planets are orbiting stars that have evolved off the main sequence into the red giant branch, and thus experience a dramatically increased incident flux compared with that received for the bulk of their life.

3.4 Binary Systems

Many stars have a stellar companion that can influence the energy received by any planets within the system. It has been well established, primarily through spectroscopic means, that stars in the solar neighborhood have a high rate of stellar multiplicity. Indeed, early studies found that binary and multiple stars systems are

[1] http://hzgallery.org/

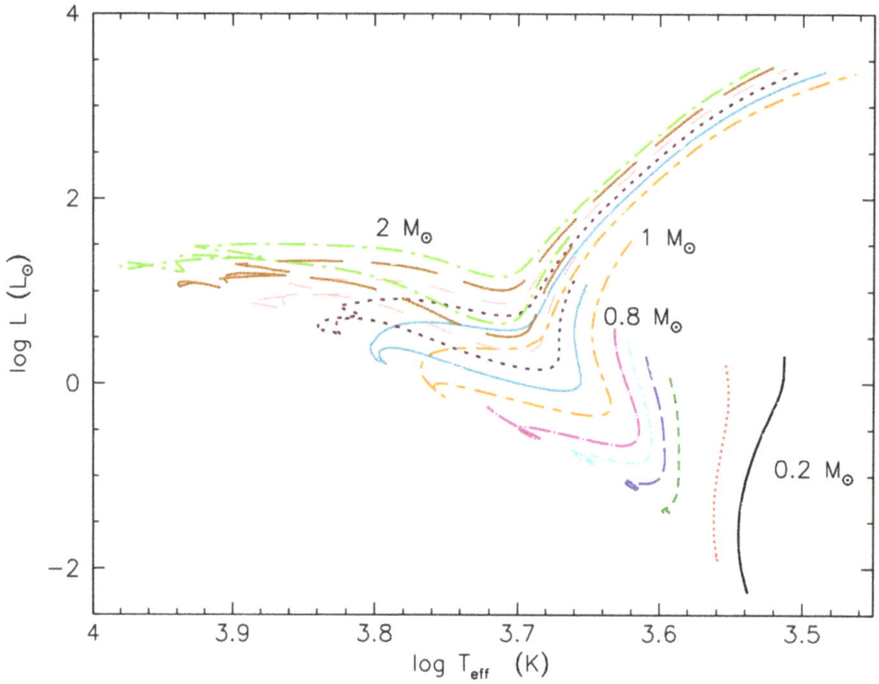

Figure 3.5. Stellar evolution pathways for stars of mass 0.2, 0.3, 0.5, 0.6, 0.7, 0.8, 1.0, 1.2, 1.4, 1.6, 1.8, and 2 $M_{\odot}$ at solar metallicity. Credit: Gallet et al. (2017), reproduced with permission © ESO.

more common than single stars (Abt & Levy 1976). More recent surveys of nearby solar-type stars have found a smaller, but still substantial, rate of stellar multiplicity (Duquennoy & Mayor 1991; Raghavan et al. 2010). These studies also found that the period distribution of solar-type binaries peaks near 10,000 days (30 years, which corresponds to about 10 AU). By contrast, observations of exoplanet-hosting binary star systems show a period distribution peak near 100 AU, higher than that is observed for field binaries and suggesting that planet formation is suppressed in close binaries (Howell et al. 2021).

For planets in binary systems, there are two basic types of orbit to consider, shown in the left panel of Figure 3.7. Planetary-type (P-type) orbits occur when a planet orbits the center of mass of a close binary, and is frequently referred to as a "circumbinary" orbit (Haghighipour & Kaltenegger 2013; Kane & Hinkel 2013; Cukier et al. 2019). Satellite-type (S-type) orbits occur when a planet orbits one stellar component of a wide binary (Quintana et al. 2007; Kaltenegger & Haghighipour 2013). Examples of both cases have been discovered. For example, two of the most eccentric known planets, HD 80606b and HD 20782b, are in S-type orbits since their host is part of a wide binary system (Naef et al. 2001; Jones et al. 2006; Laughlin et al. 2009; Kane et al. 2016). P-type orbit (circumbinary) planets may be remarkably common, depending on assumptions regarding coplanar versus an isotropic distribution of planetary orbital inclinations relative to that of their host binary star (Armstrong et al. 2014). Circumbinary planet discoveries have largely

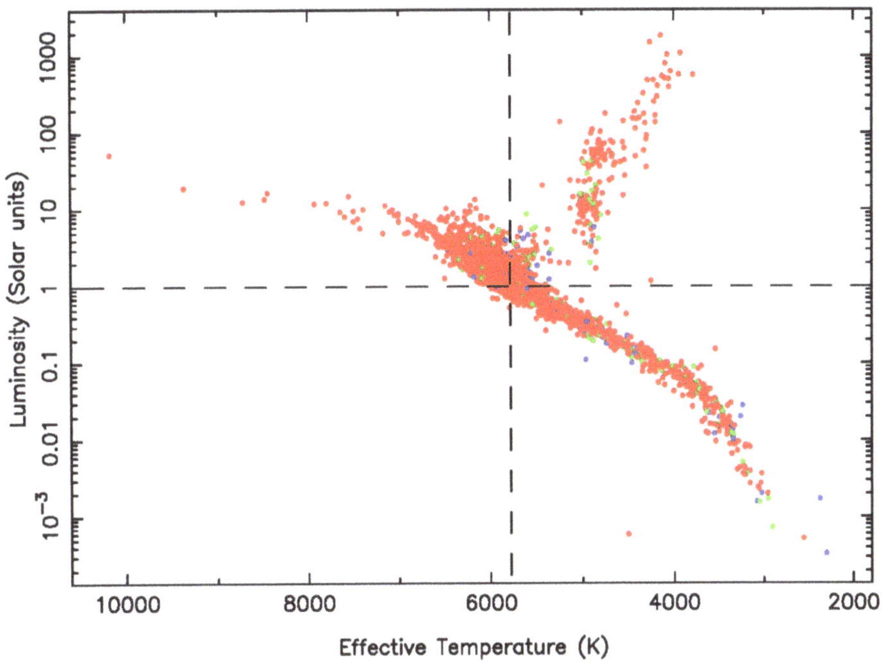

Figure 3.6. The Hertzsprung–Russell diagram (luminosity versus effective temperature) for known exoplanet hosts. The green represents planets which spend more than 50% of their orbital phase within the optimistic HZ. Red and blue represent planets interior and exterior to the HZ respectively. The dashed crosshairs in the H–R diagram show the location of the Sun. The data were extracted from the Habitable Zone Gallery (Kane & Gelino 2012).

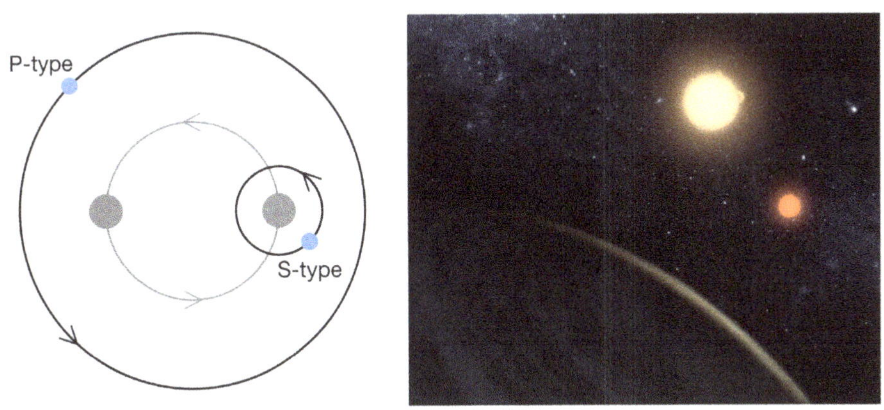

Figure 3.7. Left: a cartoon schematic of a binary star system (gray circles) containing two planets: one on a P-type (planetary-type, circumbinary) orbit and one on an S-type (satellite-type) orbit. Credit: Wikimedia: Philip D. Hall: CC BY-SA 4.0. Right: artist conception illustration of the known circumbinary planetary system, Kepler-16. Credit: Image courtesy of NASA/JPL-Caltech/T. Pyle.

stemmed from space-based transit surveys, since a planet transiting both components of a central binary produces a distinctive signature in the photometric data. The first confirmed circumbinary planetary system, discovered using data from the Kepler mission, was Kepler-16b (Doyle et al. 2011), soon to be followed by Kepler-34b & Kepler-35b (Welsh et al. 2012), and the multi-planet system Kepler-47b,c (Orosz et al. 2012). The right panel of Figure 3.7 is an artist conception illustration of the Kepler-16 system, a giant planet in a 229 day orbit around a 41 day orbit binary system.

Whereas a planet in an approximately uniform orbit around a single star will receive a relatively uniform flux at the top of the atmosphere, a planet in a circumbinary orbit can experience a significantly more complicated flux signature. Stability considerations from the three-body nature of the orbital configuration require that the planet be sufficiently removed from the center of mass of the binary (Holman & Wiegert 1999; Desidera & Barbieri 2007). For P-type orbits, this is a function of the binary separation, a_b, the mass ratio, μ, and the eccentricity of the binary orbit. The mass ratio is defined as $\mu = m_2/(m_1 + m_2)$, such that an equal mass binary has a mass ratio of $\mu = 0.5$. For a binary in a circular orbit, or $e = 0$, the "critical" radius of the planetary orbit may be expressed as

$$a_c = (1.60 + 4.12\mu - 5.09\mu^2)a_b \tag{3.4}$$

where a_c is the minimum allowed semimajor axis of the planet. The flux received by the planet will be heavily time dependent, and may be expressed as

$$F_p = \frac{L_{\star,1}}{4\pi r_1^2} + \frac{L_{\star,2}}{4\pi r_2^2} \tag{3.5}$$

where $L_{\star,1}$ and $L_{\star,2}$ are the luminosities of the two binary components, and r_1 and r_2 are the time dependent separations of the planet from each of the components.

Shown in Figure 3.8 are examples of the orbital configuration and flux received by the planet for the Kepler-34 (left) and Kepler-35 (right) systems. The top-down views shown in the top two plots show the critical semimajor axis boundary (dotted line) and planetary orbit (dashed line). Also shown as solid lines are the HZ boundaries for the systems, discussed in more detail in Chapter 6. The Kepler-34 and Kepler-35 systems are quite comparable in terms of the mass ratio and separation of the binary stars, and the similarity of the stellar components in both cases yield analogous centroids for the stability boundaries. The Kepler-34 binary consists of two near-solar components with $T_{\rm eff,1} = 5913$ K and $T_{\rm eff,2} = 5867$ K separated by 0.23 AU. The Kepler-34 planet is in an eccentric orbit with a semimajor axis of 1.09 AU. Although the flux received by the planet exhibits high frequency oscillations over one complete orbit, the total variation in incident flux is dominated by the orbital eccentricity of the planet, which periodically brings the planet closer to the binary center of mass during periastron passage (see Chapter 5). By comparison, the Kepler-35 system consists of stellar components with effective temperatures of 5606 and 5202 K, respectively, separated by 0.18 AU. As shown in the top-right panel of Figure 3.8, the planet's near-circular orbit is very close to the stability boundary of the system, and is therefore relatively close to the stellar binary. As is evident from the

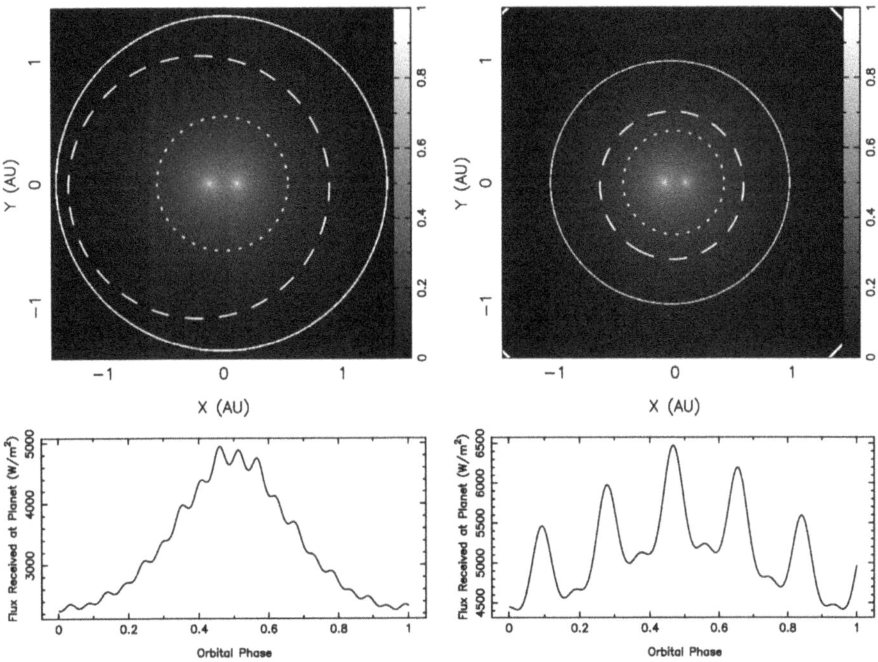

Figure 3.8. System architectures (top panels) and combined stellar flux received by the planet (bottom panels) for the Kepler-34 (left) and Kepler-35 (right) circumbinary systems (Kane & Hinkel 2013). The architecture plots show a top-down view of the systems and include the HZ boundaries (solid lines), critical semimajor axis boundary (dotted line), and planetary orbit (dashed line). The incident flux plots are presented as a function of planetary orbital phase, where phase 0.0 and 0.5 indicate the orbital locations of apastron and periastron, respectively.

bottom-right panel, this causes the incident flux profile of the planet to be very sensitive to the orbital motion of the binary.

It may be supposed that the effect of such incident flux variations in binary systems would have profound, and potentially serious, climate consequences for terrestrial planets within the HZ. However, climate simulations that are able to emulate climate forcing on timescales commensurate with those within a circumbinary system have demonstrated a climate resilience that is able to maintain long-term stability for temperate surface conditions (Popp & Eggl 2017; Komacek 2020; Wolf et al. 2020). Furthermore, tidal effects in binary systems that lower the stellar rotation rates may also lower chromospheric activity, thereby reducing atmospheric loss (see Section 3.2) and increasing sustainability of temperature environments (Mason et al. 2015). Thus, terrestrial planets in binary systems, though certainly more complicated than single star systems, may be an ideal harbor for terrestrial planet habitability.

References

Abt, H. A., & Levy, S. G. 1976, ApJS, 30, 273

Armstrong, D. J., Osborn, H. P., Brown, D. J. A., et al. 2014, MNRAS, 444, 1873

Baliunas, S. L., & Vaughan, A. H. 1985, ARA&A, 23, 379
Bessell, M. S. 1991, AJ, 101, 662
Boisse, I., Bouchy, F., Hébrard, G., et al. 2011, A&A, 528, A4
Boyajian, T. S., von Braun, K., van Belle, G., et al. 2012, ApJ, 757, 112
Butler, R. P., Marcy, G. W., Williams, E., et al. 1996, PASP, 108, 500
Campbell, B., & Walker, G. A. H. 1979, PASP, 91, 540
Charnay, B., Wolf, E. T., Marty, B., & Forget, F. 2020, SSRv, 216, 90
Chen, H., Zhan, Z., Youngblood, A., et al. 2021, NatAs, 5, 298
Ciardi, D. R., Beichman, C. A., Horch, E. P., & Howell, S. B. 2015, ApJ, 805, 16
Cukier, W., Kopparapu, R. K., Kane, S. R., et al. 2019, PASP, 131, 124402
Czesla, S., Huber, K. F., Schröter, W. S., & Schmitt, J H M M 2009, A&A, 505, 1277
Delfosse, X., Forveille, T., Perrier, C., & Mayor, M. 1998, A&A, 331, 581
Desidera, S., & Barbieri, M. 2007, A&A, 462, 345
Desort, M., Lagrange, A. M., Galland, F., Udry, S., & Mayor, M. 2007, A&A, 473, 983
Dong, C., Huang, Z., Lingam, M., et al. 2017a, ApJ, 847, L4
Dong, C., Lee, Y., Ma, Y., et al. 2018, ApJ, 859, L14
Dong, C., Lingam, M., Ma, Y., & Cohen, O. 2017b, ApJ, 837, L26
Dorn, C., Khan, A., & Heng, K. 2015, A&A, 577, A83
Doyle, L. R., Carter, J. A., Fabrycky, D. C., et al. 2011, Sci, 333, 1602
Duquennoy, A., & Mayor, M. 1991, A&A, 248, 485
Feulner, G. 2012, RvGeo, 50, RG2006
Foukal, P., Fröhlich, C., Spruit, H., & Wigley, T. M. L. 2006, Natur, 443, 161
Gaia CollaborationBrown, A. G. A., Vallenari, A., et al. 2021, A&A, 649, A1
Gallet, F., Charbonnel, C., Amard, L., et al. 2017, A&A, 597, A14
Haghighipour, N., & Kaltenegger, L. 2013, ApJ, 777, 166
Hinkel, N. R., & Unterborn, C. T. 2018, ApJ, 853, 83
Holman, M. J., & Wiegert, P. A. 1999, AJ, 117, 621
Howard, W. S., Tilley, M. A., Corbett, H., et al. 2018, ApJ, 860, L30
Howell, S. B., Matson, R. A., Ciardi, D. R., et al. 2021, AJ, 161, 164
Isaacson, H., & Fischer, D. 2010, ApJ, 725, 875
Johnson, H. L., & Morgan, W. W. 1953, ApJ, 117, 313
Jones, H. R. A., Butler, R. P., Tinney, C. G., et al. 2006, MNRAS, 369, 249
Kaltenegger, L., & Haghighipour, N. 2013, ApJ, 777, 165
Kane, S. R., & Gelino, D. M. 2012, PASP, 124, 323
Kane, S. R., & Hinkel, N. R. 2013, ApJ, 762, 7
Kane, S. R., Wittenmyer, R. A., Hinkel, N. R., et al. 2016, ApJ, 821, 65
Kiraga, M., & Stepien, K. 2007, AcA, 57, 149
Komacek, T. D. 2020, JGRE, 125, e06712
Kuiper, G. P. 1938, ApJ, 88, 472
Lammer, H., Selsis, F., Ribas, I., et al. 2003, ApJ, 598, L121
Laughlin, G., Deming, D., Langton, J., et al. 2009, Natur, 457, 562
Luger, R., & Barnes, R. 2015, AsBio, 15, 119
MacGregor, M. A., Weinberger, A. J., Wilner, D. J., Kowalski, A. F., & Cranmer, S. R. 2018, ApJ, 855, L2
Marcy, G. W., & Butler, R. P. 1992, PASP, 104, 270
Mason, P. A., Zuluaga, J. I., Cuartas-Restrepo, P. A., & Clark, J. M. 2015, IJAsB, 14, 391

Meadows, V. S., Arney, G. N., Schwieterman, E. W., et al. 2018, AsBio, 18, 133
Morgan, W. W., & Keenan, P. C. 1973, ARA&A, 11, 29
Muirhead, P. S., Hamren, K., Schlawin, E., et al. 2012, ApJ, 750, L37
Naef, D., Mayor, M., Pepe, F., et al. 2001, A&A, 375, 205
Newton, E. R., Irwin, J., Charbonneau, D., et al. 2016, ApJ, 821, L19
Newton, E. R., Mondrik, N., Irwin, J., Winters, J. G., & Charbonneau, D. 2018, AJ, 156, 217
Noyes, R. W. 1984, AdSpR, 4, 151
Orosz, J. A., Welsh, W. F., Carter, J. A., et al. 2012, Sci, 337, 1511
Owen, J. E. 2019, AREPS, 47, 67
Pace, G., & Pasquini, L. 2004, A&A, 426, 1021
Palubski, I. Z., Shields, A. L., & Deitrick, R. 2020, ApJ, 890, 30
Paulson, D. B., Cochran, W. D., & Hatzes, A. P. 2004, AJ, 127, 3579
Popp, M., & Eggl, S. 2017, NatCo, 8, 14957
Quintana, E. V., Adams, F. C., Lissauer, J. J., & Chambers, J. E. 2007, ApJ, 660, 807
Raghavan, D., McAlister, H. A., Henry, T. J., et al. 2010, ApJS, 190, 1
Ramirez, R. M., & Kaltenegger, L. 2014, ApJ, 797, L25
Rodríguez-Mozos, J. M., & Moya, A. 2019, A&A, 630, A52
Roettenbacher, R. M., & Kane, S. R. 2017, ApJ, 851, 77
Saar, S. H., & Donahue, R. A. 1997, ApJ, 485, 319
Sackmann, I. J., Boothroyd, A. I., & Kraemer, K. E. 1993, ApJ, 418, 457
Sakuraba, H., Kurokawa, H., & Genda, H. 2019, Icar, 317, 48
Santerne, A., Fressin, F., & Díaz, R. F. 2013, A&A, 557, A139
Underwood, D. R., Jones, B. W., & Sleep, P. N. 2003, IJAsB, 2, 289
Vida, K., Oláh, K., Kővári, Z., et al. 2019, ApJ, 884, 160
Welsh, W. F., Orosz, J. A., Carter, J. A., et al. 2012, Natur, 481, 475
Wilson, O. C. 1963, ApJ, 138, 832
Wolf, E. T., Haqq-Misra, J., Kopparapu, R., et al. 2020, JGRE, 125, e06576
Wright, N. J., Drake, J. J., Mamajek, E. E., & Henry, G. W. 2011, ApJ, 743, 48

Stephen R Kane

Chapter 4

Planetary Fundamentals

4.1 Constructing an Energy Balance

As described in Chapter 2, the primary challenge in the topic of planetary habitability is the construction of an energy balance at the planetary surface that properly accounts for all intervening factors. This energy balance will determine both the surface temperature and the sustainability of those conditions. Chapter 3 introduced the contribution of stars to the energy balance, and this chapter details the primary planetary processes that govern the surface environment. Like stars, these processes form one third of the "planets are hard" diagram, shown in Figure 2.1.

There are numerous planetary processes that affect the average and range of surface temperatures. For example, the transparency of the atmosphere dictates the passage of radiation received at the top of the atmosphere to the surface (see Section 4.2). The transparency is, in turn, influenced by the presence of clouds and hazes, and further modified by chemical reactions in the atmosphere. The atmosphere undergoes further chemical reactions with the surface, and the interior of the planet and state of the core results in convection of heat to the surface, in addition to volcanic outgassing. The diagram shown in Figure 4.1 summarizes the major components involved in Earth's surface energy balance, including incoming (short wave) and outgoing (long wave) radiation (Mills 2020). The radiation units are normalized to the solar radiation received at the top of the atmosphere, defined as 100 units. Accounting for scattering, reflection, and absorption effects, approximately 47% of the incoming radiation reaches the surface. The blackbody radiation emitted by the surface, which includes heat from the interior, is largely absorbed by greenhouse gases and re-emitted to the surface, with a small amount passing through infrared (IR) windows, described in Section 4.2. Note that, although the processes shown in Figure 4.1 appear static in nature, the processes involve substantial atmospheric dynamics including convection and other mechanisms that are constantly redistributing energy throughout the globe.

doi:10.1088/2514-3433/ac2aa1ch4

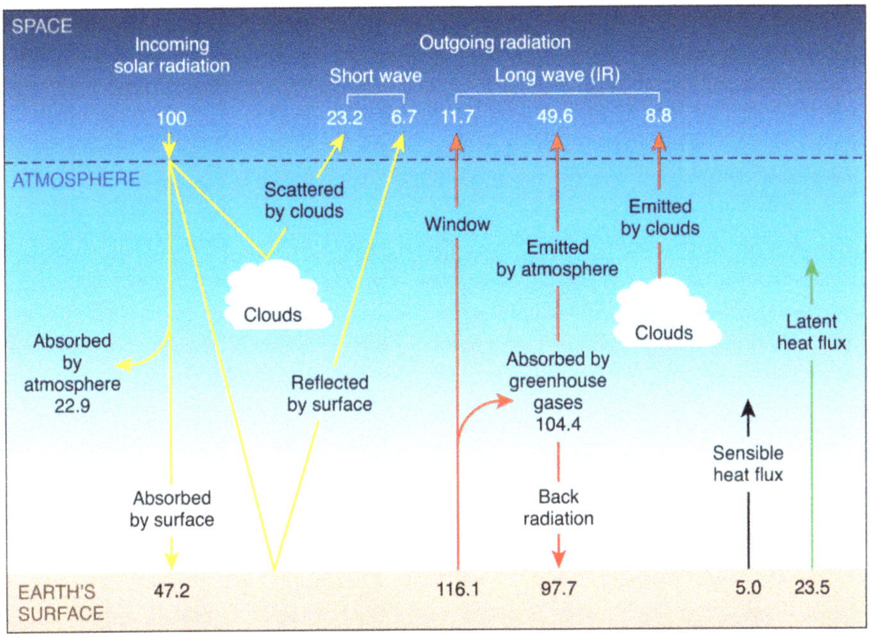

Figure 4.1. A cartoon depiction of the Earth's radiation balance (Mills 2020). The incoming short wave radiation from the Sun, defined as 100 units, is balanced by the outgoing short wave and long wave radiation. Credit: © American Meteorological Society. Used with permission.

A simple modification to the equilibrium temperature (Section 2.1) that accounts for greenhouse warming is the idealized greenhouse model. The flux received at the top of the atmosphere, F_p, is given by:

$$F_p = \frac{L_\star}{4\pi a^2}. \tag{4.1}$$

This allows Equation (2.1) to be re-expressed as:

$$T_{eq} = \left(\frac{F_p(1 - A_B)}{4\sigma}\right)^{\frac{1}{4}}. \tag{4.2}$$

For Earth, this results in an equilibrium temperature of $T_{eq} = 255$ K $= -18°$ C, not accounting for greenhouse warming. Equation (4.2) can be modified by incorporating the emissivity, ϵ, which is the ratio of the energy radiated to that radiated by a blackbody at the same temperature and wavelength. Adding a cloud (atmospheric) layer of temperature T_a to the standard blackbody model that absorbs and emits proportional to the emissivity, results in a new version of the equilibrium temperature, as follows:

$$T_s = \left(\frac{F_p(1 - A_B)}{4\sigma}\frac{1}{1 - \epsilon/2}\right)^{\frac{1}{4}}. \tag{4.3}$$

Here, T_s is the surface temperature, and is related to the atmospheric layer temperature by $T_s = 2^{1/4}T_a$. For $\epsilon = 0$, Equation (4.3) reduces to Equation (4.2), and thus yields $T_s = 255$ K. For $\epsilon = 1$, Equation (4.3) yields $T_s = 303$ K = 30° C. The actual emissivity value of $\epsilon = 0.78$ yields $T_s = 288$ K, which is close to the present mean temperature of Earth. There are numerous caveats to these calculations that are worth noting. First, the model assumes an isothermal atmosphere, whereas the temperature gradient of the atmosphere (see Section 4.2) results in more radiation emitted downward towards the surface than upward into space. Second, the model assumes the cloud layer absorbs all outgoing radiation equally at all IR wavelengths, whereas significant IR windows exist in the atmosphere, as shown in Figure 4.1 and described further in Section 4.2. Finally, the model does not account for the transfer of heat and evaporation of water, which is a significant contributor to the atmospheric energy budget.

A further aspect to consider in the context of the surface energy balance is that of the radiative equilibrium timescale (Cronin & Kerry 2013; Guendelman & Kaspi 2019). This may be thought of as the response time of the atmosphere to changes in the energy balance. Examples of such changes are variations in the incident flux from the star, or chemical/compositional changes to the atmosphere via outgassing. The seasonal changes in the surface temperature of the Earth provide a local example of the radiative equilibrium timescale, and is well-known to have a seasonal lag with respect to the insolation at the top of the atmosphere (Donohoe et al. 2020). For example, shown in Figure 4.2 is the range of seasonal variations in the ocean temperature off the coast of Protaras, Cyprus. The summer and winter solstice in the northern hemisphere occur during the date ranges of June 20–22 and December 21–22, respectively. Figure 4.2 demonstrates that, in the Mediterranean Sea, there is a lag in the minimum and maximum temperatures of ~2 months.

To estimate the radiative equilibrium timescale, τ_E, we adapt Equation (4.3) as follows:

$$\frac{1}{4}F_p(1 - A_B) + \epsilon\sigma T_a^4 = \sigma T_s^4. \tag{4.4}$$

For an atmospheric column of unit cross-sectional area in hydrostatic balance, the time dependent energy balance is given by:

$$\frac{c_P P_s}{g}\frac{dT_a}{dt} = \epsilon\sigma T_s^4 - 2\epsilon\sigma T_a^4 \tag{4.5}$$

where c_P is the heat capacity at constant pressure, P_s is the pressure at the surface, and g is the surface gravity. The mass of the atmospheric column per unit area is P_s/g. A perturbation, ΔT, to the atmospheric temperature can be described via the expression $T_a = T_{ao} + \Delta T$, where T_{ao} is the radiative equilibrium temperature of the atmosphere, and $|\Delta T| \ll T_{ao}$. To determine the time evolution of the temperature perturbation, Equations (4.4) and (4.5) may be combined to produce the following:

$$-\frac{c_P P_s}{g}\frac{d\Delta T}{dt} = \epsilon(1 - A_B)\frac{F_p}{4} + (\epsilon - 2)\epsilon\sigma(T_{ao} + \Delta T)^4. \tag{4.6}$$

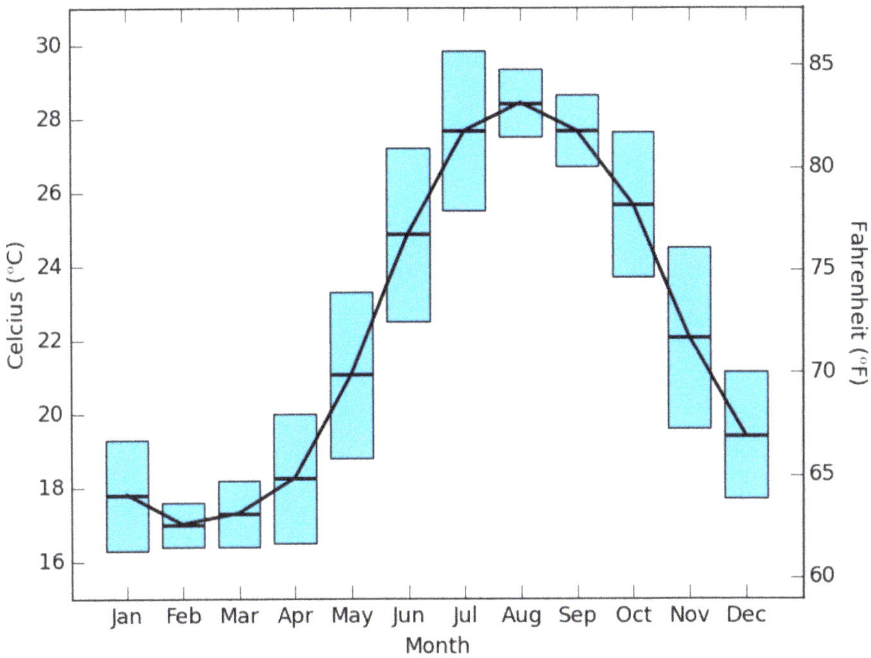

Figure 4.2. The range of annual ocean temperatures off the coast of Protaras, Cyprus, calculated from years of archival data, with a line connecting the average values. Data from the World Sea Temperatures service (http://www.seatemperature.org). Credit: Tara Fetherolf.

Equation (4.6) can be approximately solved using a Taylor series truncated after the second term:

$$f(T_{ao} + \Delta T) = f(T_{ao}) + \Delta T \frac{df(T_{ao})}{dT}. \tag{4.7}$$

Solving this results in a new expression for the time evolution of ΔT:

$$\frac{d\Delta T}{dt} = -\frac{4(2 - \epsilon)\epsilon\sigma g T_{ao}^{3}}{c_P P_s}\Delta T. \tag{4.8}$$

Integrating both sides of Equation (4.8) yields the radiative equilibrium timescale:

$$\tau_E = \frac{P_s c_P}{4(2 - \epsilon)\epsilon\sigma g T_{ao}^{3}}. \tag{4.9}$$

This represents the time required for the atmosphere to respond significantly to changes in radiative forcing and is an expression of the thermal inertia of the atmosphere.

Consider an example based on the Earth using the following parameter values: $T_{ao} = 300$ K, $\epsilon = 0.7$, $g = 10$ m s^{-1}, $P_s = 10^5$ Pa, $c_P = 1000$ J K^{-1} kg^{-1}. Applying these values to Equation (4.9) yields a radiative equilibrium timescale of $\tau_E = 20$ days. Note that Equation (4.9) is highly temperature dependent. Adopting an atmospheric temperature of $T_{ao} = 200$ K, as is the case in the upper atmosphere, yields $\tau_E = 70$ days. Furthermore, ocean water has a significantly higher heat capacity than the atmosphere, even considering the upper layers of the ocean where mixing occurs on seasonal timescales. Thus, there are a range of phase lags throughout the ocean and atmospheric layers with respect to the annual times of maximum solar radiation.

4.2 Atmospheres

A planetary atmosphere is one of the most important components to consider in the topic of planetary habitability, for two primary reasons. First, the atmosphere is the structure that provides the possibility for a sustainable temperate surface environment, modulo the energy balance requirement discussed in Section 4.1. Thus, it can be stated that an atmosphere is a necessary but insufficient requirement for planetary habitability. Second, the planetary atmosphere is one of the relatively few measurable features of an exoplanet at the present time, and often such measurements sample high elevations where opacity is relatively low. The challenge in assessing terrestrial planetary atmospheres is then to develop a pressure–temperature (P–T) profile as a function of altitude, since it is such profiles that enable the inference of surface conditions (Madhusudhan & Seager 2009). For example, the P–T profiles for solar system planets are shown in Figure 4.3, which span a variety of temperatures and pressures (Robinson & Catling 2014). There are numerous factors that influence the P–T profile of a planetary atmosphere, including (but not limited to) planet mass, radius, surface gravity, atmospheric compositional abundances, age, external

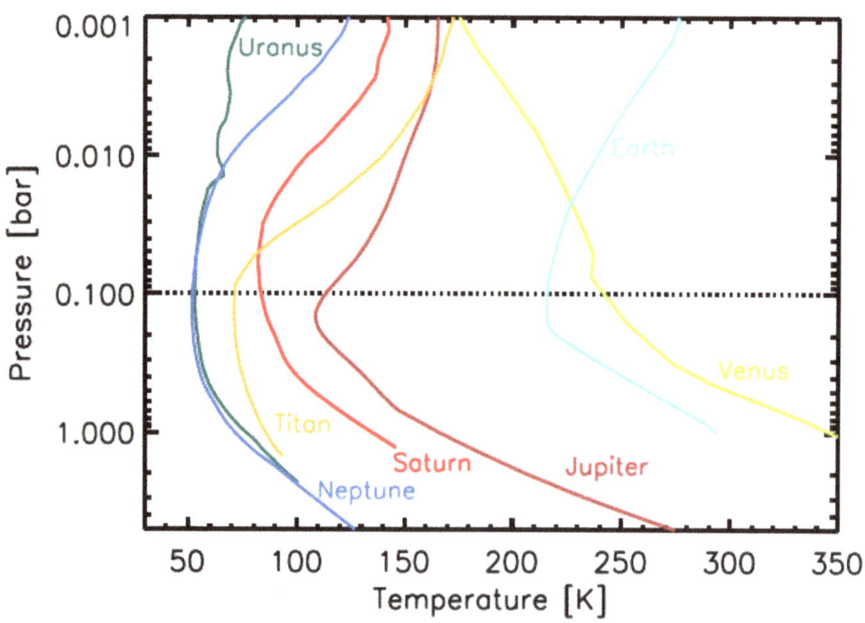

Figure 4.3. Atmospheric temperature profiles for solar system bodies with thick atmospheres. For most atmospheres, the temperature reaches a minimum at a pressure of ~0.1 bar. Credit: Reprinted by permission from Springer (GmbH): Springer Nature, Robinson & Catling (2014). © 2021 Springer Nature Limited (2014).

energy sources (star, tides), and internal energy sources (cooling, radioactivity; Seager & Deming 2010).

In general, atmospheric pressure decreases monotonically with increasing altitude. The same trend is not necessarily true of temperature, which decreases in the troposphere region of an atmosphere where convection dominates, then begins to increase at altitudes where stellar flux absorption by molecular species occurs within the stratosphere. In other words, a temperature transition occurs in a layer in which energy is released slower than it is absorbed. This "temperature inversion" is an important feature of planetary atmospheres since it represents the boundary of the deeper convective regions, and can also trap aerosols and other haze layers that truncate the atmospheric opacity (Gao et al. 2021). In Figure 4.3, it is worth noting that the temperature inversion (minimum temperature) for most planets occurs at a pressure of ~0.1 bar, thought to be a characteristic location of the tropopause for many thick atmospheres (Robinson & Catling 2014).

A standard method to assess the pressure gradient and atmospheric extent of a planet is via the scale height. The scale height of an atmosphere is the altitude over which the pressure reduces by a factor of $e \sim 2.718$. The atmospheric pressure as a function of altitude, z, may be expressed as

$$P = P_s \exp^{-z/H} \tag{4.10}$$

where P_s is the surface pressure ($z = 0$), and H is the scale height, given by

$$H = \frac{kT}{\mu m_H g}. \tag{4.11}$$

The variables/constants shown in Equation (4.11) are the Boltzmann constant, k, the temperature, T, the atmospheric mean molecular weight, μ, the mass of a hydrogen atom, m_H, and the surface gravity, $g = GM_p/R_p^2$. Notice that the scale height is inversely proportional to the surface gravity, so a more massive planet is likely to have a more dramatic reduction in pressure with altitude than a less massive planet. There are several important assumptions built in to the derivation of Equation (4.11), such as the mean molecular weight of the atmosphere being assumed to be constant as a function of pressure. However, this expression does provide a reasonable first-order estimate of the atmospheric pressure structure, particular in comparison with other bodies with different mass, radii, and surface temperatures. For example, Earth has an atmospheric scale height of ~8.5 km. The average height of Earth's troposphere, containing ~75% of the atmospheric mass, is ~13 km. By comparison, Venus, which is of similar mass and radius to Earth, has a significantly higher surface temperature and a scale height of ~15.9 km. That means it requires double the altitude on Venus to reduce the atmospheric pressure by a factor of e, compared with Earth.

A further useful characteristic of planetary atmospheres to consider is that of opacity, or optical depth, which describes the absorption and scattering of radiation as it passes through the medium of the atmosphere. The optical depth, τ_λ, is given by

$$\tau_\lambda \equiv \int_{s_1}^{s_2} \alpha_\lambda(s)\rho(s)ds \tag{4.12}$$

where α_λ and ρ are the absorption coefficient and density of the medium, respectively. The optical depth may be thought of as the number of mean free paths that occurs over a distance *ds* along the path of the radiation. For an optical depth of $\tau_\lambda \approx 1$, the intensity of the radiation is reduced by a factor of *e*. For $\tau_\lambda \gg 1$, the atmosphere is described as "optically thick" at the given wavelength, and $\tau_\lambda \ll 1$ implies the atmosphere is "optically thin" (Sharp & Burrows 2007; Heng et al. 2014). There are numerous online tools that perform opacity and radiative transfer calculations for a variety of planetary atmospheres, such as HELIOS (Malik et al. 2017; Grimm et al. 2021) and the Planetary Spectrum Generator (Villanueva et al. 2018).

The relevance of this nomenclature to planetary habitability is the role of the opacity's wavelength dependence in the radiative transfer occurring at the surface with respect to radiation received and emitted. For example, shown in Figure 4.4 is

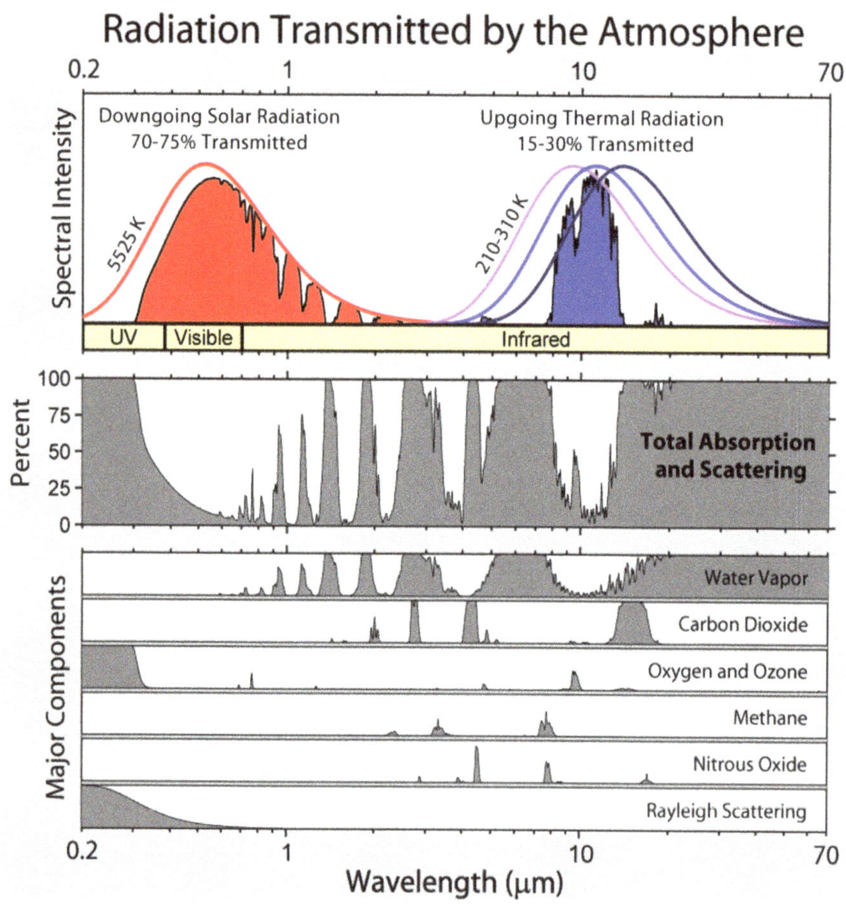

Figure 4.4. Absorption bands in the Earth's atmosphere (middle panel) and the effect on solar radiation (top panel; red) and outgoing thermal radiation (top panel; blue). Individual absorption spectrum for major greenhouse gases plus Rayleigh scattering are shown in the lower panel. Credit: Wikimedia: Robert A. Rohde: CC BY-SA 3.0.

the specific case of absorption bands in the Earth's atmosphere due to the compositional structure, along with the equivalent blackbody curves for the incoming (red) and outgoing (blue) radiation. Integrating the red and blue spectra shown in the top panel, allows the calculation of the total radiation transmitted from the top of the atmosphere to the surface and from the surface to space, respectively. The absorption bands, shown in the middle panel, are the direct result of the wavelength dependent opacity of the atmospheric species at particular temperatures and pressures and represent the optically thick wavelength regions of the atmosphere. The bottom panel shows the major species that contribute to the atmospheric absorption and scattering, including Rayleigh scattering. The presence of ozone in the atmosphere is the major absorber at ultraviolet (UV) wavelengths, preventing radiation harmful to biology from reaching the surface. The bottom panel also demonstrates that much of the overall absorption is caused by water vapor, particularly at longer wavelengths, and thus trapping much of the outgoing radiation and contributing to greenhouse warming. The high radiation absorption of water vapor also has the consequence of severely disrupting ground-based IR observations except at specific IR windows in the atmosphere. Most importantly, the atmospheric absorption is the major contributor towards the energy balance at the surface, described in Section 4.1.

4.3 Interiors

The interiors of planets plays a key role in the evolution of the surface conditions (Dehant et al. 2019). The size and composition of the planet can play a role in the cooling rate (Schubert et al. 1980; Seales & Lenardic 2021), and thus the sustainability of a magnetic dynamo (Stevenson et al. 1983) and potential crustal tectonics (Moresi & Solomatov 1998). All planets begin in a molten state as a result of the formation process that occurs within a protoplanetary disk (Lissauer 1993; Elkins-Tanton 2008; Mordasini et al. 2012). While in this molten state, the interior of the planet separates into layers via differentiation, which is caused by gravity separating different materials, where higher densities sink toward the center. This results in the formation of three distinct density zones within a terrestrial planet: core, mantle, and crust. The core tends to be metal-rich due to differentiation, and can contain significant amounts of elements such as iron and nickel. The mantle is an intermediate layer composed of silicate rock and, while considered solid, behaves as a viscous fluid over geological timescales (Dasgupta & Hirschmann 2010). The viscosity of the mantle allows heat transport from the core region to the surface through a convective process of expansion and contraction, and requires a temperature gradient between the core and the crust. The crust comprises the lowest density material and is also the top component of the lithosphere, which is the rigid, brittle material that consists of the crust and outermost part of the mantle. Smaller planets tend to have thicker lithospheres that increase in thickness as the planet cools. Shown in Figure 4.5 are schematic cross sections for the terrestrial planets of our solar system (Kane et al. 2021). These cross sections demonstrate that, even within our planetary system where all planets formed from the same protoplanetary disk, the relative sizes of the core, mantle, and crust can vary enormously. We can therefore expect that the broad population of exoplanets will have

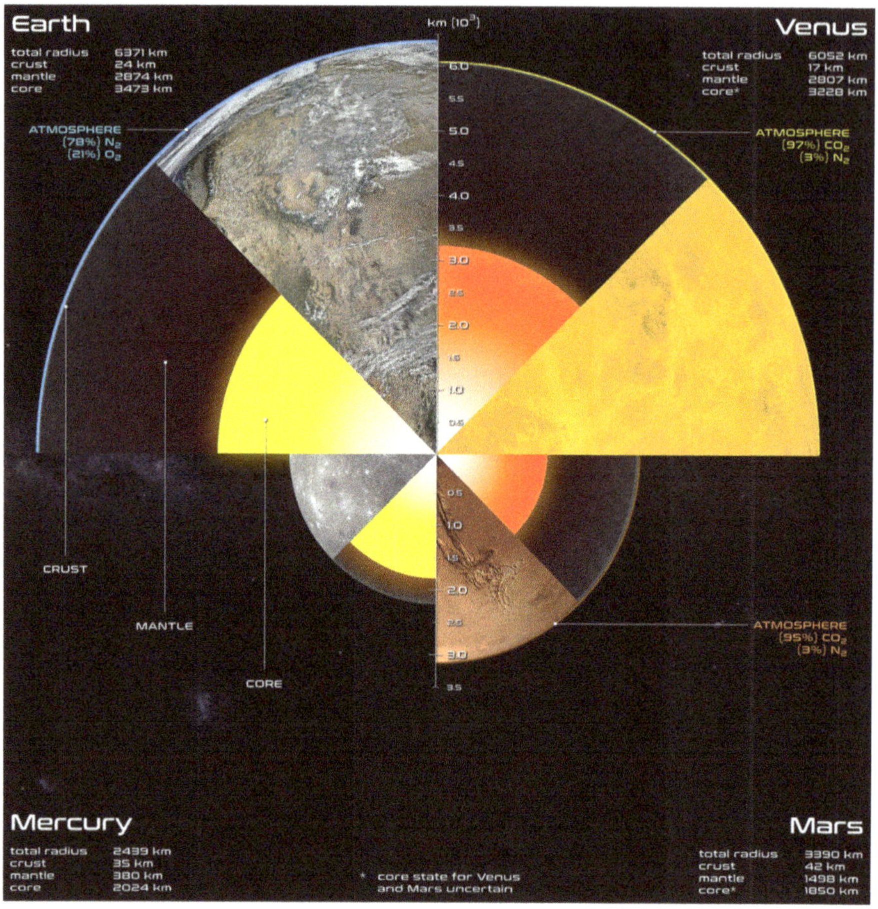

Figure 4.5. Schematic cross sections of the four inner solar system planets, showing the major internal components (crust, mantle, and core) and atmospheric components. All cross sections are to scale. For simplicity, oceanic and continental crust for Earth are not distinguished, nor is the interior structure of Earth's mantle shown. Note that there is considerable uncertainty regarding the state of the cores of Venus and Mars, and specifically whether there is a liquid and solid component, as for Earth's core, and so they are shown with orange fill instead of yellow. Credit: Kane et al. (2021), John Wiley & Sons. © 2021 American Geophysical Union.

an even greater diversity of planetary interiors (Seager et al. 2007; Elkins-Tanton & Seager 2008; Rogers & Seager 2010; Unterborn et al. 2016).

Modeling planetary interiors is based upon the available data from solar system measurements. These data largely consist of seismic and gravity measurements, which may be combined with bulk density and composition estimates (Unterborn et al. 2016). Such data are relatively sparse for solar system planets other than Earth, though seismic data for Mars (Banerdt et al. 2020) and radar data for Venus (Margot et al. 2021) are gradually improving the situation. At the fundamental level, mass and radius measurements of terrestrial planets, and the subsequent determination of the bulk density, can lead to first-order constraints that serve as boundary conditions for a more detailed analysis. As discussed in Section 2.3, masses and radii

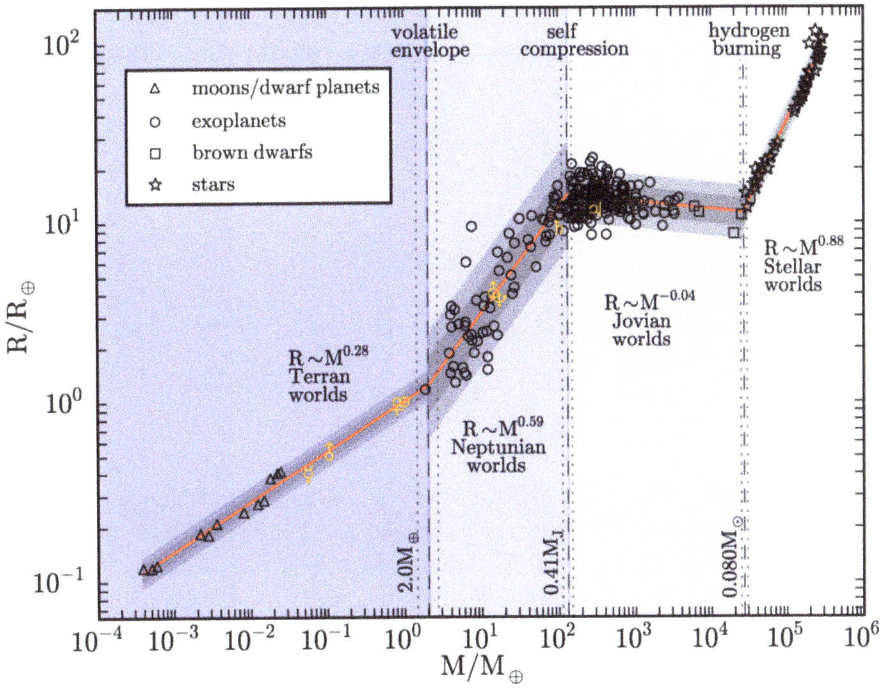

Figure 4.6. A mass–radius relationship for different classes of objects, including "Terran" (terrestrial), "Neptunian," "Jovian," and "Stellar." The vertical dashed lines indicate significant changes in the density gradients that may be representative of transitions between the different object classes. Credit: Reproduced from Chen & Kipping (2017).

may be extracted, depending on the detection technique utilized and the signal-to-noise of the measurements involved. Such extractions have led to various mass–radius relationships (Seager et al. 2007; Swift et al. 2012; Weiss & Marcy 2014), such as that shown in Figure 4.6 which demonstrates transitions between classes of objects based on their bulk densities (Chen & Kipping 2017). For terrestrial objects, a density of $\rho < 1$ g cm^{-3} implies an icy and/or porous dominated object, while $\rho > 3$ g cm^{-3} suggests an object dominated by a rocky composition (see Figure 2.3). For comparison, the Earth has a bulk density of $\rho \approx 5.5$ g cm^{-3}. In terms of bulk composition, a comparison with Earth reveals that there is a relatively limited inventory of elements that contribute to the overall interior composition. For Earth, oxygen, iron, magnesium, and silicon comprise 95% of the total mass, while including nickel, sulfur, magnesium, and calcium results in 99.9% of the mass being accounted for (Javoy 1995).

The relation between the mass and radius of a planet is primarily derived from four major principles (Sotin et al. 2010):

1. Mass conservation.
2. Energy transport.
3. Hydrostatic equilibrium.
4. Equations of state.

The first principle, conservation of mass, adopts a model of considering a shell of width dr, mass dm, and density $\rho(r)$. The mass interior to r is:

$$dm = 4\pi r^2 \rho(r) dr \tag{4.13}$$

which may be expressed as:

$$\frac{dm}{dr} = 4\pi r^2 \rho(r). \tag{4.14}$$

Integrating from the center of the planet to the surface, where $r = R_p$, results in the following expression:

$$M_p = 4\pi \int_0^{R_p} r^2 \rho(r) dr. \tag{4.15}$$

The precise nature in which the density varies with radius depends on the compositional structure of the interior. The second principle, energy transport of the interior, is controlled by the temperature profile, which may be modeled through temperature differences integrated over a particular region, such as a convective mantle. As discussed above, the temperature profile plays a major role in the amount of partial melting, and thus convection rate. This, in turn, controls the rate of energy transport to the lithosphere, whereby crustal fractures and outgassing may result, greatly influencing the surface evolution. The third principle, hydrostatic equilibrium, is the balance between gravity and pressure that determines the relationship between temperature, pressure, and density. The change in pressure, dP, across a shell of width dr is given by:

$$dP = -g\rho(r) dr \tag{4.16}$$

where $g = Gm/r^2$ is the gravity at radius r. This results in the expression:

$$\frac{dP}{dr} = -\frac{Gm}{r^2}\rho(r). \tag{4.17}$$

Assuming a uniform density, and using the boundary condition of zero pressure at the surface ($P = 0$ where $r = R_p$), provides an expression for the pressure at the center of the planet:

$$P_c = \frac{3GM_p^2}{8\pi R_p^4}. \tag{4.18}$$

The consequence of the uniform density assumption is an underestimate of the true central pressure. For example, the expression yields a central pressure for Earth of ~172 GPa, whereas the true central pressure is ~330 GPa (Alfè et al. 2007), a factor of two higher. A physically motivated density profile may be achieved through the fourth principle, which utilizes equations of state. The equations of state account for phases of materials in a planet as a function of temperature, pressure, and density. The methodology therefore assumes functions $P = P(\rho, T)$ and are often expressed

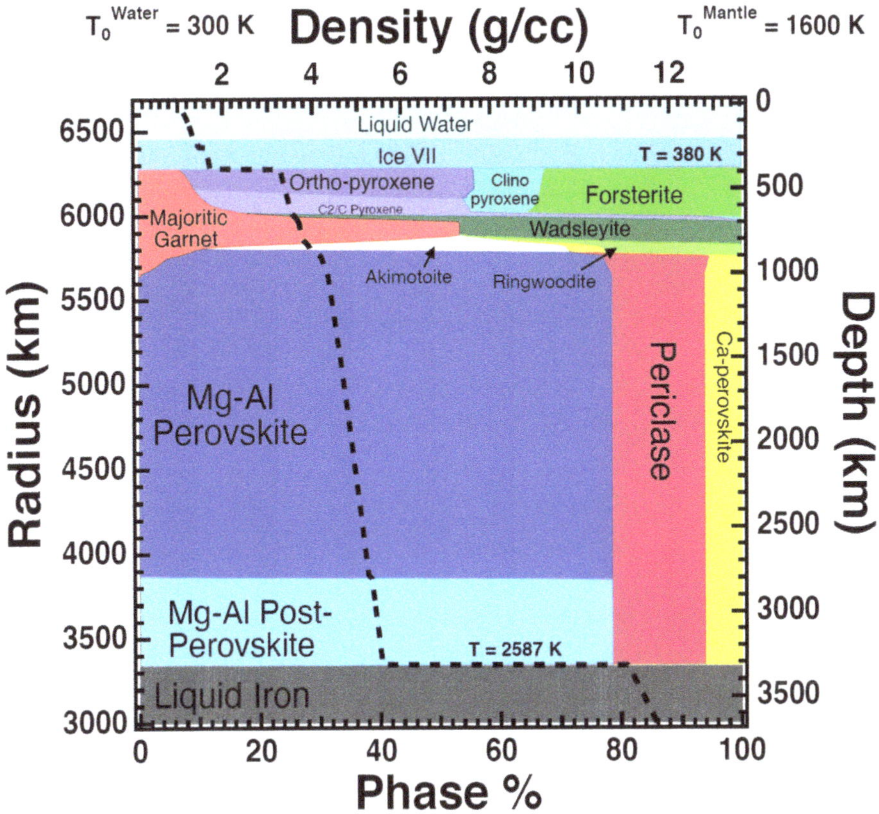

Figure 4.7. A phase diagram for an interior model of the exoplanet TRAPPIST-1f (Unterborn et al. 2018). Adiabatic temperatures are shown at each major boundary, and the dashed line represents the change in density. Credit: Reprinted with permission from Cayman Unterborn.

as a series of polytropes of the form $P = K\rho^{(n+1)/n}$, where K and n are the polytropic constant and index, respectively. For example, Figure 4.7 shows the resulting phase diagram from an interior model of the available data for the exoplanet TRAPPIST-1f (Unterborn et al. 2018). The model assumes a water-rich composition based on bulk density measurements, implying an inward migration of the planet after formation beyond the "snow line," which is the distance from a star where volatile compounds (such as water) are able to condense from the nebula during formation and contribute substantially to planetary compositional inventories. Adiabatic temperatures are shown at the major boundaries, and the density profile is shown as a dashed line. Such models are generally data limited, particularly regarding compositional information. This is especially true for exoplanets, where stellar elemental abundances may be used to infer refractory abundances that comprise the bulk of the Earth's composition (Hinkel & Unterborn 2018).

As noted earlier, the convective processes within a planet may result in plate tectonics within the lithosphere (O'Neill et al. 2007). Although plate tectonics is well established as a feature of Earth's geology, the mechanisms by which it operates is an active area of research (Moresi & Solomatov 1998; Tackley 2000; Foley & Driscoll 2016), and its occurrence on Venus remains a topic of continuous discussion (Smrekar et al. 2018). Furthermore, it is unknown how the occurrence of plate tectonics scales with terrestrial planet mass (Foley et al. 2012), since the higher interior pressures may act to inhibit (O'Neill & Lenardic 2007) or promote (Valencia et al. 2007) mantle convection, depending on assumptions regarding shear stresses with respect to plate motion. The critical importance of this topic is due to its key role in the carbonate-silicate cycle that is observed on Earth, and the temperature-dependent "thermostat" effect that enables long-term sustainability of temperate surface conditions (Walker et al. 1981; Berner et al. 1983; Schaefer & Sasselov 2015; Oosterloo et al. 2021). Thus, a deeper understanding of the relationship between energy transport, mantle convection, plate tectonics, and atmospheric carbon cycling is a key component for addressing sustainable planetary habitability (Menou 2015; Rushby et al. 2018).

4.4 Planetary Evolution

The atmospheres and interiors of planets undergo significant changes as the planet ages (McGovern & Schubert 1989; Kadoya & Tajika 2015). A terrestrial planet's primary atmosphere originates from nebular material in an accretion disk. The primary atmosphere mainly consists of H, H_2, and He and may also contain trace elements, such as CO_2, CH_4, N_2, H_2O and NH_3. Such atmospheres may be relatively short-lived, particularly for planets with low surface gravity and/or high surface temperatures. The combination of these factors with the influence of the host star (see Section 3.2) can effectively desiccate the primary atmosphere (Howe et al. 2020). For terrestrial planets, most of the H will escape, leaving heavier gases, such as argon, neon, and ammonia, concentrated near the surface. The primary atmosphere is replaced with the secondary atmosphere, by which gases trapped in planetesimals during formation are outgassed from the interior as differentiation causes them to rise to the surface of the planet (Kite & Barnett 2020). Earth's primitive atmosphere contained H_2, H_2O, CO, and H_2S, and volcanic gases account for most of the present Earth's secondary atmosphere (Sagan & Mullen 1972; Kasting 1993). Figure 4.8 shows a model for the CO_2 outgassing of terrestrial planets as a function of time, for planetary masses of 0.2 $M_\oplus$ (dotted line), 1.0 $M_\oplus$ (solid line), and 5.0 $M_\oplus$ (dashed line; Kadoya & Tajika 2015). The luminosity evolution of a solar-mass star is also shown (see Section 3.3), which has implications for both the atmospheric erosion and energy balance of the planet (see Sections 3.2 and 4.1). Further contributions to the secondary atmosphere can originate from comets and meteors that contain water and gas (H_2O, CH_4, CO_2; Avice et al. 2017; Marty et al. 2017; Bekaert et al. 2020). Although outgassing and cometary contributions to the secondary atmosphere generally decline with time, biological contributions are more stochastic in nature, and can produce significant atmospheric changes when the biomass crosses a critical threshold (Lyons et al. 2014; Avice et al. 2017).

In the context of planetary evolution, it is worth revisiting the issue of atmospheric erosion (escape), that was briefly described in Section 3.2. Escape mechanisms fall into the two broad categories of thermal and non-thermal processes

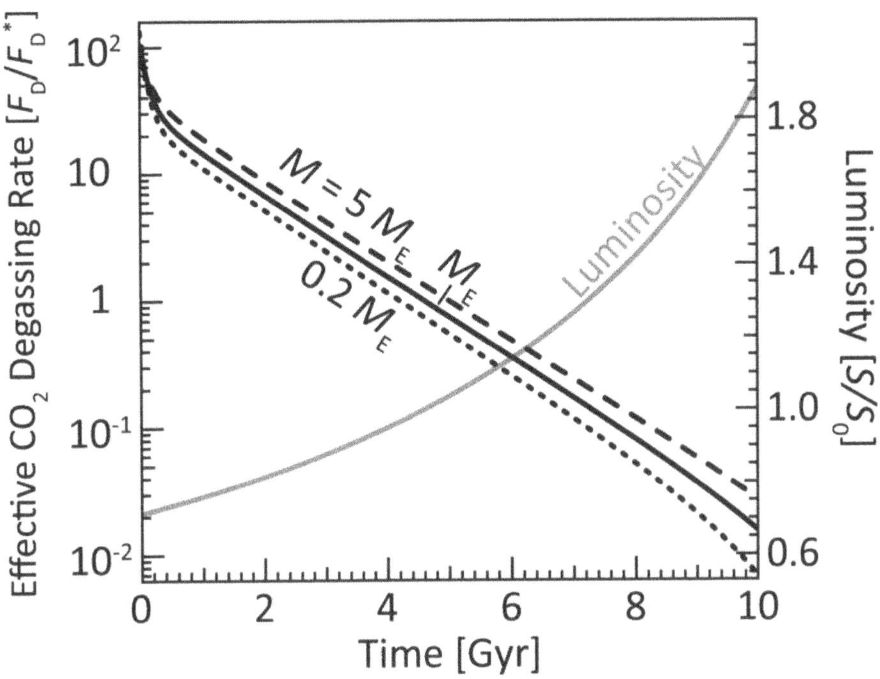

Figure 4.8. Outgassing of CO_2 for terrestrial planets as a function of time. The planetary masses included are 0.2 $M_\oplus$ (dotted line), 1.0 $M_\oplus$ (solid line), and 5.0 $M_\oplus$ (dashed line). Also shown is the luminosity evolution of a solar-mass star. Credit: Reproduced from Kadoya & Tajika (2015).

(Lammer et al. 2008; Volkov et al. 2011; Dong et al. 2018). Additional atmospheric escape may occur via giant impacts that largely take place during a planet's early history when the primary atmosphere is still dominant (Schlichting et al. 2015; Kane et al. 2020). Thermal escape processes, such as Jeans escape and hydrodynamic escape, are generally described in terms of the kinetic energy of molecules with respect to the planetary gravity. The escape velocity of a planet is given by:

$$v_{\rm esc} = \sqrt{\frac{2GM_p}{R_p}} \tag{4.19}$$

where M_p and R_p are the planetary mass and radius, respectively. A molecule of mass m and temperature T will have an rms velocity of:

$$v_{\rm rms} = \sqrt{\frac{3kT}{m}}. \tag{4.20}$$

One may consider that atmospheric escape will occur when $v_{\rm rms} > v_{\rm esc}$. However, this assumes that a single mean free path of a molecule is required to escape. Collisions result in a velocity distribution of molecules in the form of a Maxwell distribution, which contains a high-velocity tail where atmospheric escape tends to occur.

Significant molecular escape then requires that $6v_{rms} > v_{esc}$, which can be written in the form of a temperature requirement:

$$T > \frac{1}{54}\frac{GM_p}{R_p}\frac{m}{k}. \tag{4.21}$$

Applying Equation (4.21) to several case studies within the solar system demonstrates that Jupiter is more than capable of retaining hydrogen molecules, whereas Neptune's moon Triton, despite the cold surface temperatures, cannot retain a hydrogen atmosphere. Non-thermal escape generally involve some combination of photochemical processes and interactions with charged particles (Shizgal & Arkos 1996). For example, the upper atmospheres of terrestrial planets are often vulnerable to UV radiation from the host star, whose high energy can photodissociate molecules and provide additional energy that enables the smaller components to escape. Escape of the resulting ionized particles may be mitigated by the presence of a substantial magnetic field (Stevenson et al. 1983; Owen & Adams 2019), though the overall beneficial aspects of the interactions between magnetic field and escaping ions remains uncertain (Gunell et al. 2018). The Martian atmosphere has long been suspected of being a victim to the solar wind (Levine et al. 1974; Brain & Jakosky

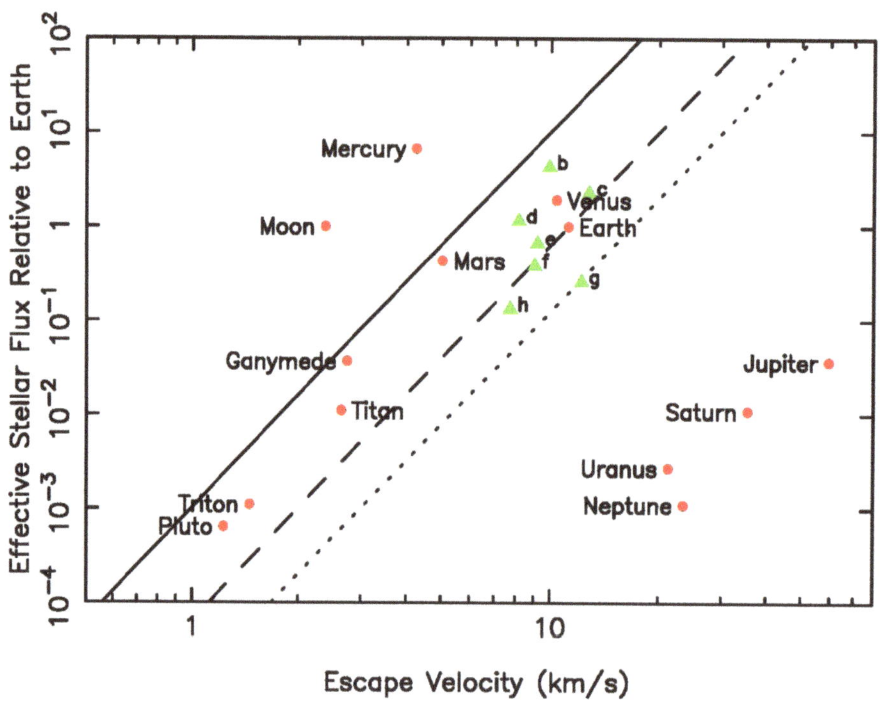

Figure 4.9. Effective stellar flux incident upon the planet as a function of the calculated escape velocity for solar system objects (red circles) and the TRAPPIST-1 planets (green triangles). Two additional lines for the cosmic shoreline downweight the escape velocity by a factor of 2 (dashed) and 4 (dotted) to incorporate the effect of larger XUV flux compared with solar values. Credit: Reproduced from Roettenbacher and Kane (2017).

1998; Chassefière & Leblanc 2004). Indeed, observations by the Mars Atmosphere and Volatile Evolution (MAVEN) mission (Jakosky et al. 2015) have revealed that atmospheric mass loss has played the dominant role in the surface evolution of Mars (Jakosky et al. 2018; see Section 7.4). Shown in Figure 4.9 is a representation of atmospheric loss using the incident stellar flux received by a planet and the planetary escape velocity (Roettenbacher & Kane 2017). The solid line is an empirically determined location of the "cosmic shoreline" (Zahnle & Catling 2017), based upon the observational evidence of atmospheric retention for objects within the solar system (red circles). Objects above the line have negligible atmosphere, while those below the line retain a substantial atmospheric pressure. The dashed and dotted lines incorporate the increased XUV emission of TRAPPIST-1 relative to the Sun, which can therefore increase the expected non-thermal escape of the atmospheres for the TRAPPIST-1 planets (green triangles). As described in Section 3.2, the activity of low-mass stars may thus present a significant inhibitor to the long-term habitability of terrestrial planets.

4.5 A Note on Water Worlds

The premise of planetary habitability described in Chapter 2 is that of conditions that lead to a sustainable presence of surface liquid water. This has led to considerable interest in icy moons within the solar system, whose formation beyond the snow line resulted in water-rich interiors (see Section 7.6). More broadly, water delivery and formation mechanisms have led to substantial speculation regarding the prevalence of so-called "water worlds," defined as terrestrial planets, and even mini-Neptunes, with high relative water inventories and that lack continents (Léger et al. 2004; Fu et al. 2010; Tian & Ida 2015). The detection of water vapor in the atmosphere of a planet may be an indicator that the planet is indeed water-rich (Komacek et al. 2020; Smith et al. 2020); and thus is a highly sought after signature in the field of exoplanet spectroscopy (Benneke et al. 2019).

However, the water premise of planetary habitability may lead to a possible misconception that habitability scales monotonically with the fractional water content of a planet. There are several caveats to note in this regard. First, the majority of a planet's water inventory may be present in the interior rather than the surface. For example, the water content of Earth's mantle is thought to exceed the total surface volume (Bell & Rossman 1992; Hirschmann & Kohlstedt 2012), and may in fact be substantially larger (Peslier et al. 2017). This means that estimates of bulk water content for a planetary body may not be indicative of surface hydration (Cowan & Abbot 2014). Second, models of water temperature and pressure environments have demonstrated that a sufficiently deep ocean can result in a high-pressure ice layer at the crust–mantle boundary, thereby truncating geochemical processes between the surface and the mantle (Noack et al. 2016; Glaser et al. 2020). As discussed in Section 4.3, the combination of continental weathering and oceanic crustal subduction provide an important means to sequester atmospheric carbon into the interior, and thus water worlds may have limited capacity to engage in temperature-dependent modulation of atmospheric CO_2 (Marounina & Rogers 2020). Furthermore, cycling of phosphorus, an essential bioelement (Bjerrum & Canfield 2002; Hinkel et al. 2020), is expected to be dramatically impeded by high-pressure ice layers at the ocean floors of water worlds, potentially imposing serious constraints for the progression of biochemistry (Glaser et al. 2020). It is therefore possible that there is an optimal surface water inventory that maximizes the potential for sustainable planetary habitability.

References

Alfè, D., Gillan, M. J., & Price, G. D. 2007, ConPh, 48, 63
Avice, G., Marty, B., & Burgess, R. 2017, NatCo, 8, 15455
Banerdt, W. B., Smrekar, S. E., Banfield, D., et al. 2020, NatGe, 13, 183
Bekaert, D. V., Broadley, M. W., & Marty, B. 2020, NatSR, 10, 5796
Bell, D. R., & Rossman, G. R. 1992, Sci, 255, 1391
Benneke, B., Wong, I., Piaulet, C., et al. 2019, ApJL, 887, L14
Berner, R. A., Lasaga, A. C., & Garrels, R. M. 1983, AmJS, 283, 641
Bjerrum, C. J., & Canfield, D. E. 2002, Natur, 417, 159
Brain, D. A., & Jakosky, B. M. 1998, JGR, 103, 22689
Chassefière, E., & Leblanc, F. 2004, P&SS, 52, 1039
Chen, J., & Kipping, D. 2017, ApJ, 834, 17
Cowan, N. B., & Abbot, D. S. 2014, ApJ, 781, 27
Cronin, T. W., & Emanuel, K. A. 2013, JAMES, 5, 843
Dasgupta, R., & Hirschmann, M. M. 2010, E&PSL, 298, 1
Dehant, V., Debaille, V., Dobos, F., et al. 2019, SSRv, 215, 42
Dong, C., Jin, M., Lingam, M., et al. 2018, PNAS, 115, 260
Donohoe, A., Dawson, E., McMurdie, L., Battisti, D. S., & Rhines, A. 2020, JCli, 33, 3921
Elkins-Tanton, L. T. 2008, E&PSL, 271, 181
Elkins-Tanton, L. T., & Seager, S. 2008, ApJ, 685, 1237
Foley, B. J., Bercovici, D., & Landuyt, W. 2012, E&PSL, 331, 281
Foley, B. J., & Driscoll, P. E. 2016, GGG, 17, 1885
Fu, R., O'Connell, R. J., & Dimitar, 2010, ApJ, 708, 1326
Gao, P., Wakeford, H. R., Moran, S. E., & Parmentier, V. 2021, JGRE, 126, e06655
Glaser, D. M., Hartnett, H. E., Desch, S. J., et al. 2020, ApJ, 893, 163
Grimm, S. L., Malik, M., Kitzmann, D., et al. 2021, ApJS, 253, 30
Guendelman, I., & Kaspi, Y. 2019, ApJ, 881, 67
Gunell, H., Maggiolo, R., Nilsson, H., et al. 2018, A&A, 614, L3
Heng, K., Mendonça, J. M., & Lee, J.-M. 2014, ApJS, 215, 4
Hinkel, N. R., Hartnett, H. E., & Young, P. A. 2020, ApJL, 900, L38
Hinkel, N. R., & Unterborn, C. T. 2018, ApJ, 853, 83
Hirschmann, M., & Kohlstedt, D. 2012, PhT, 65, 40
Howe, A. R., Adams, F. C., & Michael, 2020, ApJ, 894, 130
Jakosky, B. M., Brain, D., Chaffin, M., et al. 2018, Icar, 315, 146
Jakosky, B. M., Lin, R. P., Grebowsky, J. M., et al. 2015, SSRv, 195, 3
Javoy, M. 1995, GeoRL, 22, 2219
Kadoya, S., & Tajika, E. 2015, ApJ, 815, L7
Kane, S. R., Arney, G. N., Byrne, P. K., et al. 2021, JGRE, 126, e06643
Kane, S. R., Rottenbacher, R. M., Unterborn, C. T., et al. 2020, PSJ, 1, 36
Kasting, J. F. 1993, Sci, 259, 920

Kite, E. S., & Barnett, M. N. 2020, PNAS, 117, 18264
Komacek, T. D., Fauchez, T. J., Wolf, E. T., & Abbot, D. S. 2020, ApJ, 888, L20
Lammer, H., Kasting, J. F., Chassefiere, E., et al. 2008, SSRv, 139, 399
Léger, A., Selsis, F., Sotin, C., et al. 2004, Icar, 169, 499
Levine, J. S., Keating, G. M., & Prior, E. J. 1974, P&SS, 22, 500
Lissauer, J. J. 1993, ARA&A, 31, 129
Lyons, T. W., Reinhard, C. T., & Planavsky, N. J. 2014, Natur, 506, 307
Madhusudhan, N., & S Seager, 2009, ApJ, 707, 24
Malik, M., Grosheintz, L., Mendonca, J. M., et al. 2017, AJ, 153, 56
Margot, J.-L., Campbell, D. B., Giorgini, J. D., et al. 2021, NatAs, 5, 676
Marounina, N., & Rogers, L. A. 2020, ApJ, 890, 107
Marty, B., Altwegg, K., Balsiger, H., et al. 2017, Sci., 356, 1069
McGovern, P. J., & Schubert, G. 1989, E&PSL, 96, 27
Menou, K. 2015, E&PSL, 429, 20
Mills, E. W. 2020, Weather Studies: Introduction to Atmospheric Science (Boston, MA: American Meteorological Society)
Mordasini, C., Alibert, Y., Klahr, H., & Henning, T. 2012, A&A, 547, A111
Moresi, L., & Solomatov, V. 1998, GeoJI, 133, 669
Noack, L., Höning, D., Rivoldini, A., et al. 2016, Icar, 277, 215
O'Neill, C., Jellinek, A. M., & Lenardic, A. 2007, E&PSL, 261, 20
O'Neill, C., & Lenardic, A. 2007, GeoRL, 34, L19204
Oosterloo, M., Höning, D., Kamp, I. E. E., & van der Tak, F. F. S. 2021, A&A, 649, A15
Owen, J. E., & Adams, F. C. 2019, MNRAS, 490, 15
Peslier, A. H., Schönbächler, M., Busemann, H., & Karato, S.-I. 2017, SSRv, 212, 743
Robinson, T. D., & Catling, D. C. 2014, NatGe, 7, 12
Roettenbacher, R. M., & Kane, S. R. 2017, ApJ, 851, 77
Rogers, L. A., & Seager, S. 2010, ApJ, 712, 974
Rushby, A. J., Johnson, M., Mills, B. J. W., Watson, A. J., & Claire, M. W. 2018, AsBio, 18, 469
Sagan, C., & Mullen, G. 1972, Sci, 177, 52
Schaefer, L., & Sasselov, D. 2015, ApJ, 801, 40
Schlichting, H. E., Sari, R., & Yalinewich, A. 2015, Icar, 247, 81
Schubert, G., Stevenson, D. J., & Cassen, P. 1980, JGR, 85, 2531
Seager, S., Kuchner, M., Hier-Majumder, C. A., & Militzer, B. 2007, ApJ, 669, 1279
Seager, S., & Deming, D. 2010, ARA&A, 48, 631
Seales, J., & Lenardic, A. 2021, Icar, 367, 114560
Sharp, C. M., & Burrows, A. 2007, ApJS, 168, 140
Shizgal, B. D., & Arkos, G. G. 1996, RvGeo, 34, 483
Smith, A. J. R. W., Feng, Y. K., Fortney, J. J., et al. 2020, AJ, 159, 36
Smrekar, S. E., Davaille, A., & Sotin, C. 2018, SSRv, 214, 88
Sotin, C., Jackson, J. M., & Seager, S. 2010, Terrestrial Planet Interiors (Tucson, AZ: Univ. Arizona Press), 375
Stevenson, D. J., Spohn, T., & Schubert, G. 1983, Icar, 54, 466
Swift, D. C., Eggert, J. H., Hicks, D. G., et al. 2012, ApJ, 744, 59
Tackley, P. J. 2000, Sci, 288, 2002
Tian, F., & Ida, S. 2015, NatGe, 8, 177
Unterborn, C. T., Dismukes, E. E., & Panero, W. R. 2016, ApJ, 819, 32

Unterborn, C. T., Desch, S. J., Hinkel, N. R., & Lorenzo, A. 2018, NatAs, 2, 297
Valencia, D., O'Connell, R. J., & Sasselov, D. D. 2007, ApJ, 670, L45
Villanueva, G. L., Smith, M. D., Protopapa, S., Faggi, S., & Mandell, A. M. 2018, JQSRT, 217, 86
Volkov, A. N., Johnson, R. E., Tucker, O. J., & Erwin, J. T. 2011, ApJ, 729, L24
Walker, J. C. G., Hays, P. B., & Kasting, J. F. 1981, JGR, 86, 9776
Weiss, L. M., & Marcy, G. W. 2014, ApJ, 783, L6
Zahnle, K. J., & Catling, D. C. 2017, ApJ, 843, 122

Stephen R Kane

Chapter 5

Orbit, Spin, and System Effects

5.1 Keplerian Orbits

The final third of the "planets are hard" diagram (Figure 2.1) that has not yet been addressed is that which pertains to the orbital properties of the planet and overall system dynamics. The essential importance of the system dynamics is divided into two primary aspects. First, a full description of the orbit and orientation of the planet is required to calculate the time-dependent flux received at the top of the atmosphere and as a function of the location on the planet. Second, interactions of the planet with other bodies within the system may induce significant tidal energy. Both of these aspects are important considerations in the calculation of the overall surface energy balance, described in Section 4.1. Furthermore, exoplanet discoveries have revealed a large diversity of planetary architectures (Ford 2014; Winn & Fabrycky 2015; Hatzes 2016; He et al. 2019), particularly planets whose orbital eccentricities are far larger than those found in the solar system (Shen & Turner 2008; Kane et al. 2012). For example, Figure 5.1 shows a top-down planetary system view, where the dotted crosshairs indicate the location of the star (Kane et al. 2021). The orbits of the four solar system terrestrial planets are shown in black, where Mercury has the largest orbital eccentricity of any planet within the solar system. Shown in red are the orbits for the exoplanets HD 20782b (Kane et al. 2016), HD 80606b (Bonomo et al. 2017), and HD 37605b (Wang et al. 2012), whose orbital eccentricities far exceed that of Mercury.

The star–planet separation at any time is governed by the motion of the planet through a Keplerian orbit. Briefly, Kepler's laws are as follows:

1. The orbit of a planet around the Sun is an ellipse with the Sun at one focus.
2. As a planet moves around in its orbit, it sweeps out equal areas in equal times.
3. The orbital period, P, and semimajor axis, a, of a planetary orbit are related by $P^2 = a^3$, where P is in units of years and a is in units of astronomical units, AU.

It may seem rather remarkable that Kepler's third law is such a simple relationship. However, since the expression is based on both Earth's orbital period and separation

doi:10.1088/2514-3433/ac2aa1ch5

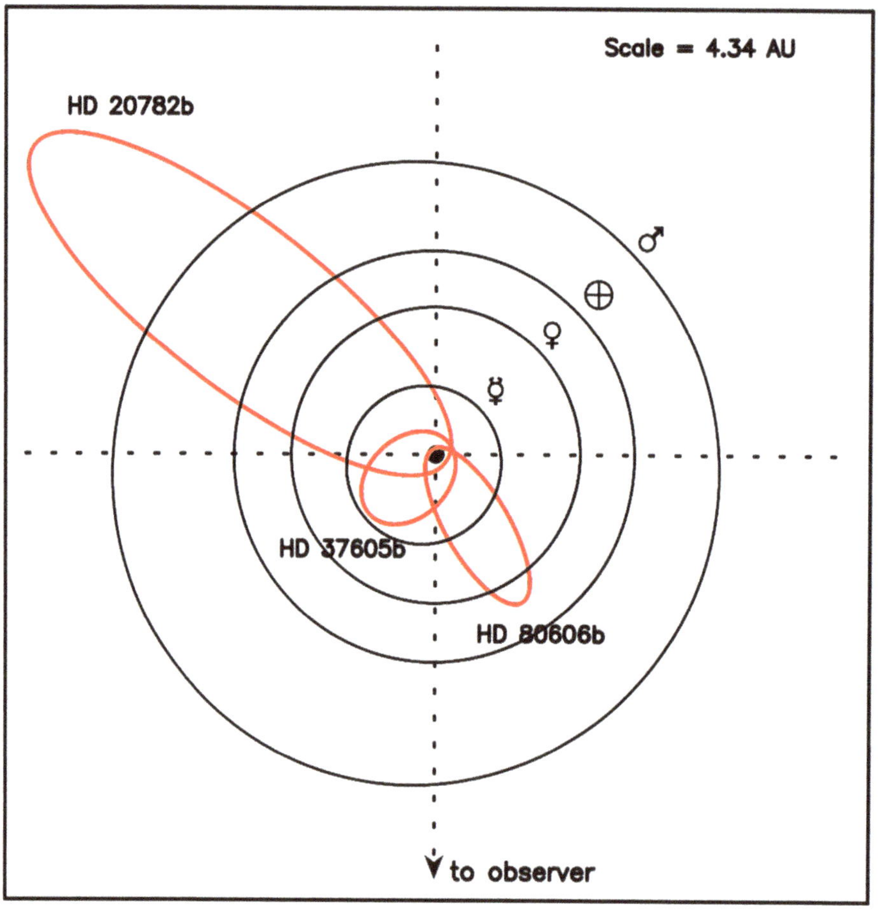

Figure 5.1. A top-down planetary system view with the star in the center, the direction to the observer, and the scale along each side of the box is 4.34 AU. Shown in red are the orbits for three exoplanets with relatively high orbital eccentricities. For comparison, the orbits of the four solar system terrestrial planets are shown in black. Credit: Kane et al. (2021). John Wiley & Sons. © 2021. The Authors.

from the Sun, the expression is essentially normalized to the specific case of Earth's orbit. More generally, we use Newton's version of Kepler's third law to apply the relationship to any stellar mass, $M_\star$, and planet mass, M_p, as follows:

$$P^2 = \frac{4\pi}{G(M_\star + M_p)} a^3 \tag{5.1}$$

where SI units are applied. Note that, particularly for terrestrial planets, $M_\star \gg M_p$. Therefore, Equation (5.1) may be expressed in more convenient units as follows:

$$P = \frac{365.256 a^{3/2}}{M_\star^{1/2}} \tag{5.2}$$

or alternatively, for the semimajor axis, as follows:

$$a = \frac{M_\star^{1/3} P^{2/3}}{51.097} \tag{5.3}$$

where P is in days, a is in AU, and $M_\star$ is in solar masses.

However, of primary concern in the context of habitability are Kepler's first and second laws, which describe the changing star–planet separation. The left panel of Figure 5.2 shows four example orbits, ranging from eccentricities of $e = 0.0$ (circular) to $e > 1.0$ (hyperbolic). Bound orbits are those for which the eccentricity is less than 1.0. Notice that the central body (star) is located at the focus denoted by F, in accordance with Kepler's first law. The geometry of an eccentric orbit, shown in red, is depicted in the right panel of Figure 5.2. The center of the ellipse is denoted by C, the time-dependent location of the planet is denoted P, and, once again, the location of the star is at the focus F. The semimajor axis, a, is the radius of the auxiliary circle, shown in blue. The location of the closest approach of the planet to the star is called the periastron, denoted A. The farthest distance of the planet from the star is referred to as the apastron (not shown). The semimajor axis is often mistakenly cited as the mean separation of the star and planet. However, because of Kepler's second law, the semimajor axis is correctly defined as the average of the periastron and apastron distances from the star (Williams 2003).

There are three time-dependent angles (called anomalies) that describe the angular location of the planet in its orbit. The true anomaly, $f(t)$, is the angle between the periastron direction and planet position, measured from the ellipse barycenter (angle PFA). The eccentric anomaly, $E(t)$, is the angle between the periastron direction and the planet position projected from the ellipse semimajor axis onto the auxiliary circle, measured from the ellipse center (angle $P'CA$).

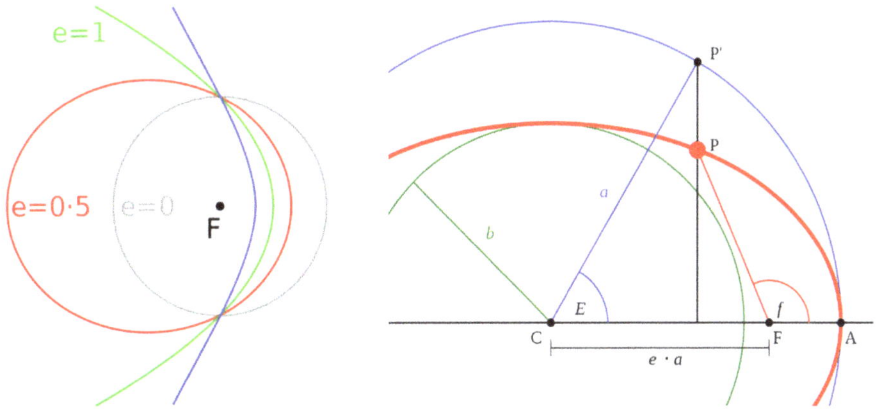

Figure 5.2. Left: depiction of various orbital eccentricities, ranging from circular (gray) to hyperbolic (blue). The location of the central body (star) is denoted by F. Right: geometry of an ellipse, shown as a thick red line. The semimajor axis, a, is the radius of the auxiliary circle, shown in blue. Also, shown are the eccentric anomaly, E, the true anomaly, f, relative to the star at focus F and planet at location P. Credit: Wikimedia: (left) ScottAlanHill: CC BY-SA 3.0, (right) CheChe: CC BY-SA 4.0.

The mean anomaly, $M(t)$, not shown on the figure, represents a fictitious mean motion around the orbit, given by:

$$M(t) = \frac{2\pi}{P}(t - t_p) \tag{5.4}$$

where $2\pi/P$ is the mean angular rate of motion, and t_p is the time at periastron passage. $M(t)$ and $E(t)$ are related by:

$$M(t) = E(t) - e \sin E(t). \tag{5.5}$$

This is referred to as Kepler's equation. The transcendental nature of Kepler's equation means that it has no analytic solution, and must therefore be solved numerically or graphically. By calculating $M(t)$ and solving Kepler's equation, the true anomaly may be extracted via the following relation:

$$\cos f(t) = \frac{\cos E(t) - e}{1 - e \cos E(t)}. \tag{5.6}$$

The star–planet separation at any time t is then calculated from the following:

$$r(t) = \frac{a(1 - e^2)}{1 + e \cos f(t)}. \tag{5.7}$$

Thus, for any Keplerian orbit, it is the values of r and f that provide the polar coordinates of the planet. Consider also the boundary conditions of periastron and apastron. At periastron, $f = 0°$ and Equation (5.7) reduces to $r = a(1 - e)$. Similarly, at apastron, $f = 180°$ and Equation (5.7) reduces to $r = a(1 + e)$. It is worth noting two other angles that are particularly useful when describing an orbit from the perspective of an external observer, and are therefore frequently used in the context of exoplanet observations. First, the argument of periastron, ω, is the angle between the periastron and the ascending node in the orbital plane. For exoplanets, ω is then essentially the angle between periastron and the plane of the sky. Second, the inclination of the orbit with respect to the plane of the sky, i, varies between 0–90°, where $i = 0°$ is an orbit observed face-on, and $i = 90°$ is an orbit observed edge-on. The inclination is particularly important for transit planets, whose observation requires that the orbit be such that $i \sim 90°$.

With the time-dependence of the planetary location now quantified, Equation (4.1) may be rewritten as:

$$F_p = \frac{L_\star}{4\pi r(t)^2} \tag{5.8}$$

which now represents the time-dependent flux. It has been demonstrated that the orbital eccentricity can have significant consequences for the climate evolution of terrestrial planets (Williams & Pollard 2002; Dressing et al. 2010; Kane & Gelino 2012; Bolmont et al. 2016; Méndez & Rivera-Valentín 2017; Way & Georgakarakos 2017; Palubski et al. 2020). Although, the maximum flux occurs during periastron passage, Kepler's second law dictates that the planet will also be traveling the fastest at that location. Depending on the extent of the eccentricity, this may result in a "flash heating" of the planetary atmosphere, as has been observed for eccentric giant planets, such as HD 80606b (Laughlin et al. 2009). On the other hand, the apastron passage can be far longer in duration and, for apastron locations relatively far from the star, the energy balance at the surface may depend on the thermal inertia of the surface environment to maintain temperate conditions. Indeed, it is worth noting that the temperature modulations due to changes in flux at the top of the atmosphere can be dampened by the thermal inertia of the planetary surface and atmosphere, such as substantial liquid water oceans (see Section 4.1). Furthermore, measurements of eccentricities for small planets in compact systems have shown that they tend to have low eccentricities (Kane et al. 2012; Van Eylen & Albrecht 2015), although this has yet to be demonstrated for long-period terrestrial planets, particularly those under the perturbing influence of a giant planet.

5.2 Rotation and Obliquity

Although the star–planet separation described in Section 5.1 allows the calculation of the flux received at the top of the atmosphere, the spin-state (rotation) and orientation (obliquity) of the planet determine the distribution of that flux over the sphere. Rotation rate determines the rate at which energy is delivered to a particular location on a planet, and can result in substantial changes to the energy balance, as well as condensation cycles and silicate weathering (Jansen et al. 2019). Shown in Figure 5.3 are the results of climate modeling of the present Earth that changes the rotation rate (Guzewich et al. 2020). These simulations show that the transition from a fast rotator to a slow rotator or tidally locked state results in increased cloud formation at the substellar point. This in turn increases the planetary albedo which has a cooling effect on the planet. Such studies are particularly important since it is expected that many planets are expected to be in a tidally locked or low spin–orbit resonance (Barnes 2017). In some cases, climate simulations of tidally locked planets have revealed conditions that lead to climate instability (Kite et al. 2011) and potential atmospheric collapse (Wordsworth 2015; Auclair-Desrotour & Heng 2020). Others have studied the relationship between atmospheric circulation, convection, and thermal structure that tidal locking produces (Koll & Abbot 2016; Sergeev et al. 2020; Hammond & Lewis 2021). The combination of exoplanet spectra with the climate data available from the solar system will be crucial in robustly determining the dependence of climate dynamics and energy balance on the rotational state (Kane et al. 2021).

The obliquity is defined as the angle between the planet's rotational axis and a line through the center of the planet perpendicular to the orbital plane. The obliquity of a planet's rotational axis may undergo oscillations of various timescales due to perturbations from other planetary bodies in the system (Laskar & Robutel 1993; see Section 5.3). These obliquity oscillations can lead to significant effects on climates (Williams & James 1997; Heller et al. 2011; Kane & Torres 2017) and extreme obliquities can move the outer edge of the habitable zone (HZ; Williams &

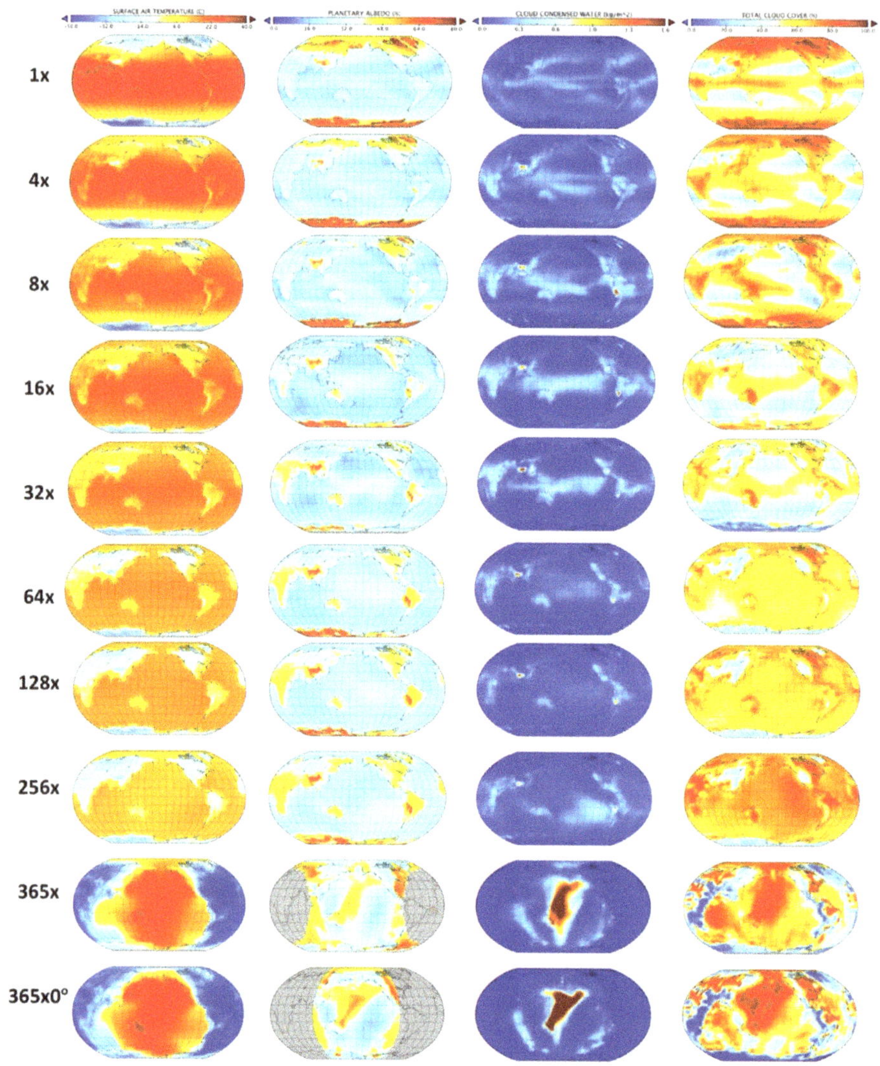

Figure 5.3. The effect of rotation rate, shown as a multiplier of Earth's present rotation rate, on (from left to right) surface air temperature, planetary albedo, column-integrated cloud-condensed water, and total cloud cover. Credit: Reproduced from Guzewich et al. (2020).

Pollard 2003; Armstrong et al. 2014; Linsenmeier et al. 2015; Colose et al. 2019). Expanding upon Equation (5.8), the maximum stellar flux received by a planet at latitude β is given by

$$\begin{aligned} F &= \frac{L_\star}{4\pi r^2}(\sin\delta\sin\beta + \cos\delta\cos\beta) \\ &= \frac{L_\star}{4\pi r^2}\cos|\beta - \delta| \end{aligned} \tag{5.9}$$

where $L_\star$ is the stellar luminosity (Kane & Torres 2017). This maximum flux can be considered the instantaneous flux, or the latitudinal flux received as a planet nears synchronous rotation (tidal locking). The solar declination, δ, is given by:

$$\delta = \theta \cos[2\pi(\phi - \Delta\phi)] \tag{5.10}$$

where ϕ is the orbital phase, $\Delta\phi$ is the offset in orbital phase between periastron and highest solar declination in the northern hemisphere, and θ is the obliquity. For Earth, $\Delta\phi = 0.46$ and $\theta = 23.5°$. The top panel of Figure 5.4 is an incident flux map for an Earth–Sun analog as a function of latitude with contours of constant flux throughout a complete orbital phase. The phase of $\phi = 0.0$ corresponds to the planet's periastron passage.

For planets where the rotation period is significantly less than the orbital period, the average incident flux as a function of latitude may be used. For this purpose, Equation (5.9) is modified as follows:

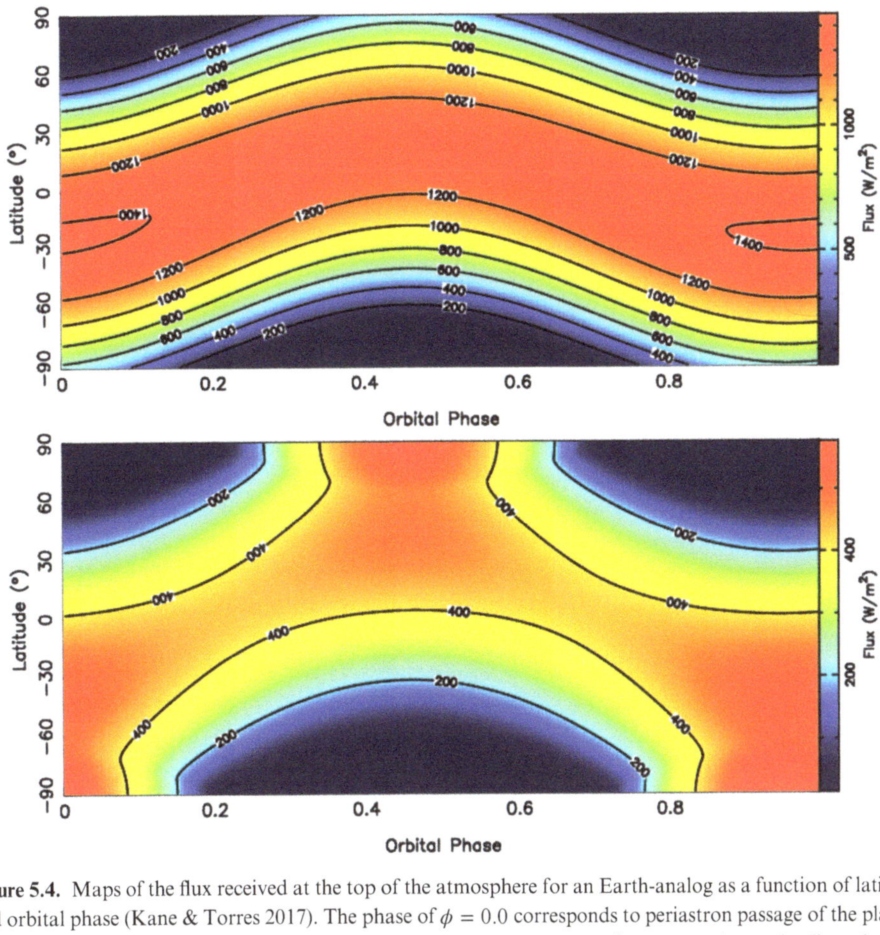

Figure 5.4. Maps of the flux received at the top of the atmosphere for an Earth-analog as a function of latitude and orbital phase (Kane & Torres 2017). The phase of $\phi = 0.0$ corresponds to periastron passage of the planet. Top: maximum (instantaneous) flux applicable to a slow rotator. Bottom: flux averaged over the diurnal cycle, application to a rapid rotator.

$$F = \frac{L_\star}{4\pi r^2}(\sin\delta \sin\beta + \cos\delta \cos\beta \cos h) \tag{5.11}$$

where h is the hour angle of the host star with respect to the local meridian. The fraction of planetary rotation period that experiences daylight for a given latitude is:

$$\Delta t_{dl} = \frac{2\arccos(-\tan\delta \tan\beta)}{360^\circ}. \tag{5.12}$$

For obliquities of $\theta > 0^\circ$, there will be latitudes that experience constant day/night during the course of an orbital period. These situations are defined by the criteria that if $\beta + \delta > 90^\circ$ or $\beta + \delta < -90^\circ$ then $\Delta t_{dl} = 1.0$, and if $\beta - \delta > 90^\circ$ or $\beta - \delta < -90^\circ$ then $\Delta t_{dl} = 0.0$. The average flux at a given latitude can then be calculated by accounting for the change in flux as a function of h and the fractional daylight time. The bottom panel of Figure 5.4 is an incident flux map averaged over the diurnal cycle for an Earth–Sun analog as a function of latitude. The comparison with the top panel of Figure 5.4 shows the impact of including the effect of constant daylight periods on the polar regions.

5.3 Planet–Planet Interactions

The discussion thus far in this chapter has focused on an individual planet that is orbiting the host star, unperturbed by other factors. However, planets are always gravitationally perturbing each other in ways that, integrated over time, can lead to non-negligible alterations in their respective energy budgets (Laskar et al. 2004). The orbital changes can result in cyclic variations in the orbital eccentricity, obliquity, and the axial orientation with respect to the celestial sphere (precession) which manifests as precession of the equinoxes (Rosengren & Scheeres 2014). These are referred to as Milankovitch cycles (Berger 1980) and have been studied extensively with regards to Earth's dynamical and climate history (Olsen & Kent 1999; Lisiecki 2014; Horner et al. 2020). Shown in Figure 5.5 are some of the cycles for Earth, including precession, obliquity, and eccentricity data (Quinn et al. 1991), as well as solar forcing (insolation flux) and glaciation data (Lisiecki & Raymo 2005), where the gray bars indicate interglacial periods. These cycles are caused by the transfer of angular momentum between Earth and other planets in the solar system, primarily Venus, Jupiter, and Saturn (Laskar 1994). As noted in Section 5.2; the current obliquity of Earth is $\theta \sim 23.5^\circ$. Milankovitch cycles result in a variation of obliquity between 22.1° and 24.5° over a ~41,000 year timescale. The variations on the eccentricity of Earth's orbit are also relatively mild. The current eccentricity is $e = 0.017$ and varies in the range 0.000 055–0.067 9 with a primary timescale of ~413,000 years. The calculation of Milankovitch cycles have also been applied to exoplanetary systems (Spiegel et al. 2010; Deitrick et al. 2018a, 2018b; Horner et al. 2020), including terrestrial planets in binary star systems (Forgan 2016).

The detailed methodology of determining orbital stability in multi-planet systems, including the solar system, is through perturbation theory, primarily developed by Joseph-Louis Lagrange and Pierre-Simon Laplace. Calculating the motion of planets within a system and correctly taking into account all of their respective

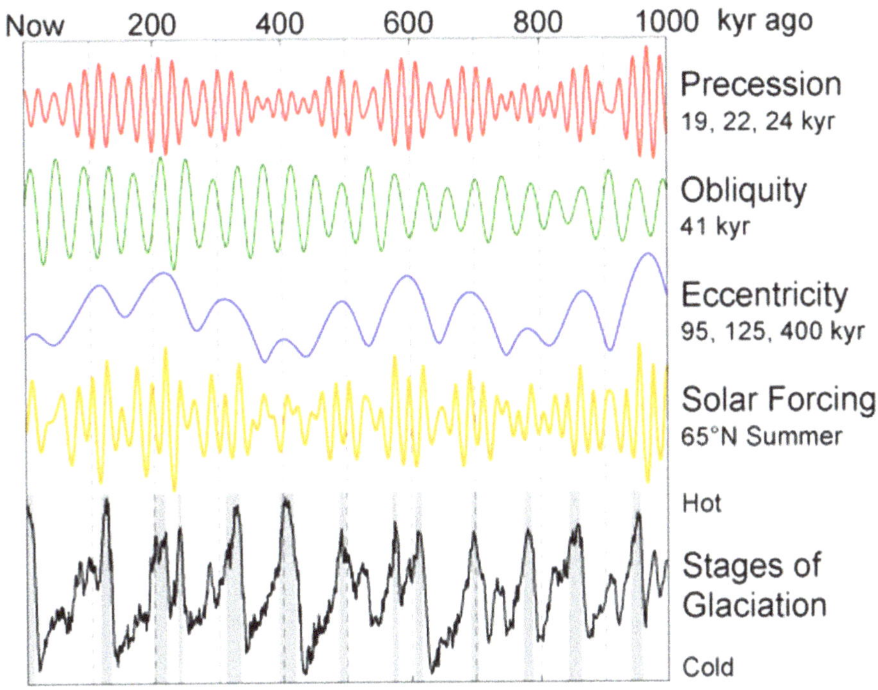

Figure 5.5. Variations in the precession, obliquity, eccentricity, solar forcing, and glaciation of the Earth over the past million years. Credit: Wikimedia: Robert A. Rohde: CC BY-SA 3.0.

perturbations is a computationally expensive process (Rasio & Ford 1996; Chatterjee et al. 2008). Most of the approaches to these calculations adopt a symplectic integrator, which is a numerical integration scheme specifically designed for systems governed by Hamiltonian physics (Kinoshita et al. 1991; Saha & Tremaine 1992; Duncan et al. 1998; Rein & Tamayo 2015). There are several publicly available N-body integrators in common usage, such as Mercury (Chambers 1999) and REBOUND (Rein & Liu 2012).

As described in Section 5.1, there is a vast diversity of planetary orbital eccentricities, many of which are in stark contrast to the near-circular orbits we find within the solar system. Planets in high eccentricity orbits contain more angular momentum than their circular counterparts (assuming similar semimajor axis and planetary mass), and that angular moment is transferred to other planets within the system, often on relatively short timescales. For example, shown in Figure 5.6 are the predicted eccentricity evolutions of two terrestrial planets in the Kepler-1649 system, whose size and insolation flux are similar to Venus and Earth (Kane et al. 2021). Notice that the predicted amplitude of the eccentricity oscillations ($e \sim 0.2$) exceeds the highest planetary eccentricity in the solar system (Mercury) where, for Earth-like obliquities, the seasons would be driven by the star–planet separation rather than the obliquity (see Section 5.2). Furthermore, the period of the oscillations is only ~2,500 years; much shorter than Earth's Milankovitch cycles. Such a continuously variable eccentricity may have significant climate stabilization

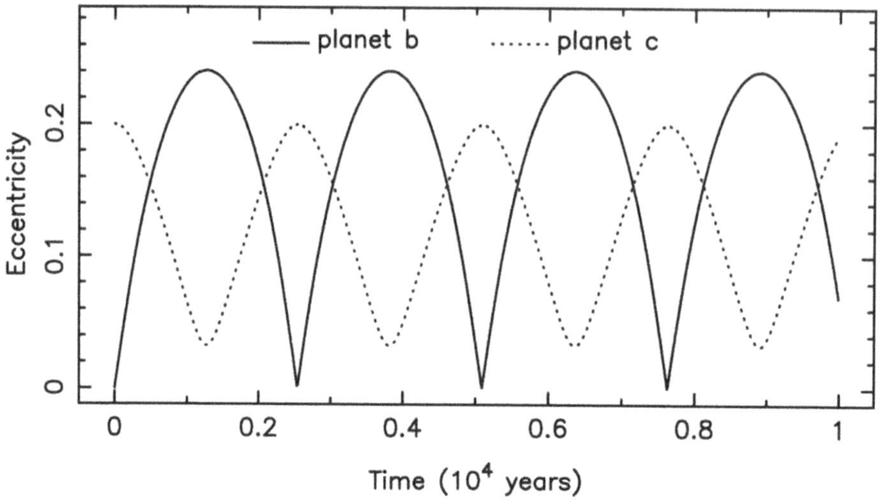

Figure 5.6. Predicted eccentricity as a function of time for two terrestrial planets in the Kepler-1649 system. Credit: Reproduced from Kane et al. (2021).

implications, and result in regular periods of glaciation during apastron (Way & Georgakarakos 2017).

Planetary systems that are presently observed to retain substantial orbital eccentricities may indicate past planet–planet scattering events that dramatically changed the architecture of the system (Carrera et al. 2019; Ford & Rasio 2008; Jurić & Tremaine 2008). Although scenarios in which planets form in eccentric orbits have been investigated (Clement et al. 2021a, 2021b); another possibility is a significant transfer of angular momentum to an additional component (stellar or planetary) no longer present in the system (Kane & Raymond 2014). To first order, this can be quantified through the use of the angular momentum deficit (AMD), which is the difference in angular momentum of the system compared with the angular momentum if the planets were in circular orbits with the same semimajor axes. This is given by the following equation:

$$\mathrm{AMD} \equiv \sum_{i=1}^{N} L_p\left(1 - \sqrt{1 - e_i^2}\right) \tag{5.13}$$

where

$$L_p = \frac{M_{p,i}M_\star}{M_{p,i} + M_\star}\sqrt{G(M_{p,i} + M_\star)a_i} \tag{5.14}$$

is the angular momentum of a circular orbit for planet i. Note that in cases where $M_{p,1} \sim M_{p,2}$, the inner planet has much less orbital angular momentum than the outer planet resulting in larger eccentricity oscillations for a given AMD value.

Of particular interest are dynamical perturbations that influence terrestrial planets within the HZ of their stars (Kane 2015; Agnew et al. 2019). Dynamical simulations have been used to predict stable locations where as-yet undiscovered HZ terrestrial planets may be found (Kopparapu et al. 2009; Kopparapu & Barnes 2010). This may be especially relevant to compact planetary systems that harbor more than one planet in the HZ, such as the TRAPPIST-1 system (Gillon et al. 2017; Luger et al. 2017). A generalized approach to the problem is to investigate the dynamical limitations of how many planets may reside within a system (Barnes & Quinn 2004; Smith & Lissauer 2009, 2010; Obertas et al. 2017). For example, Figure 5.7 shows the results of an exhaustive dynamical simulation that tested the limits of how many terrestrial planets may be packed within the HZ of a system while retaining dynamical stability (Kane et al. 2020). The simulation showed that, for most stellar types, five planets can maintain long-term stability in the HZ, whereas increasing that number to six or seven can result in dynamical instability, depending on the host star and the location of orbital resonances. For the solar system, the simulations show that six terrestrial planets are able to remain stable in the HZ, but the seven planet architecture falls within the regime of catastrophic instability.

However, the simulation results shown in Figure 5.7 neglect the effect of giant planets that can also produce perturbative effects that can disrupt dynamical stability (Lunine 2001; Georgakarakos et al. 2018). Terrestrial planet formation is substantially influenced by the presence of giant planets, particularly during their early periods of migration (Morbidelli & Raymond 2016). It is widely established that the solar system giant planets have played a significant role in the formation, evolution, and final architecture of the inner solar system (Tsiganis et al. 2005; Batygin & Laughlin 2015; Horner et al. 2020), and that such important dynamical effects of giant planets have similarly shaped the architecture of other systems (Matsumura et al. 2013; Morbidelli & Raymond 2016). For example, a planet within the asteroid belt of the solar system would significantly alter the orbital dynamics of

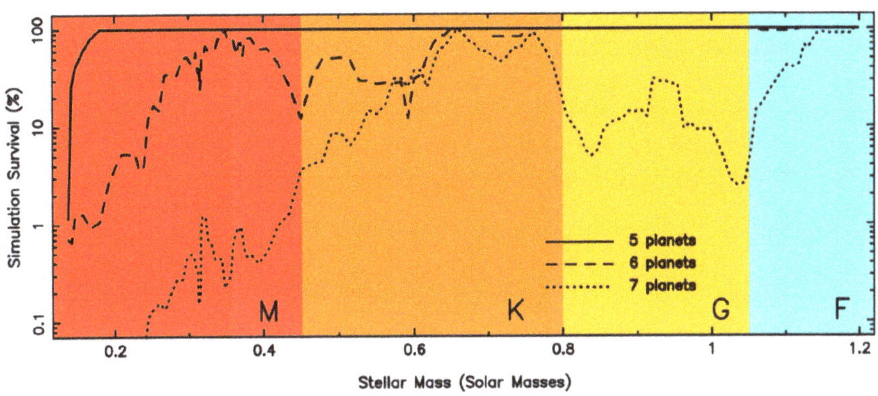

Figure 5.7. Results of a simulation that tests the dynamical survival of five, six, and seven Earth-mass planets that are evenly spaced within the HZ, as a function of the mass of the host star. The cases of five, six, and seven planets are shown as solid, dashed lines, and dotted lines respectively. The colored backgrounds indicate the corresponding spectral types (see bottom right of each colored region) for the stellar mass ranges shown. Credit: Reproduced from Kane et al. (2020).

the inner solar system by exchanging angular momentum with the outer solar system (Lissauer et al. 2001). It has been suggested from observational data that giant planets are relatively rare around M dwarfs (Endl et al. 2006; Cumming et al. 2008; Johnson et al. 2010; Bonfils et al. 2013) and that this may influence the habitability of terrestrial planets in those systems (Horner & Jones 2008; Shields et al. 2016; Kane & Blunt 2019). Even though giant planets are more common around solar analogs than later-type stars (Zechmeister et al. 2013; Wittenmyer et al. 2016; Fulton et al. 2021), they are still relatively rare, with an average occurrence rate of 6.7% beyond orbital periods of 300 days (Wittenmyer et al. 2020). Although our habitability case-study of Earth resides within a system that harbors several giant planets, the overall giant planet occurrence rate results suggest that the dynamical influence of giant planets on planetary habitability may be statistically negligible.

5.4 Moons and Exomoons

There has been substantial work conducted regarding the detection of exomoons and their potential habitability (Reynolds et al. 1987; Williams et al. 1997; Heller 2012; Heller & Barnes 2013; Hinkel & Kane 2013; Forgan & Yotov 2014; Heller et al. 2014; Kipping et al. 2015). In the scenarios of a terrestrial planet harboring a moon, or the terrestrial body itself being a moon of a giant planet, there are additional contributions and modifications to the energy balance that should be considered. The Earth–Moon system is an example of the former scenario, and it has been commonly postulated that the presence of a moon with substantial mass played a key role in Earth's habitability. For example, related to the discussion in Section 5.2 and 5.3; calculations of evolution in Earth's spin angular momentum vector indicate that a moonless Earth may experience extreme variations in obliquity, resulting in dramatic climate changes (Laskar et al. 1993; Tomasella et al. 1996; Lissauer et al. 2012; Li & Batygin 2014). However, the necessity of a substantial moon toward planetary habitability remains unclear, and indeed it is unlikely that terrestrial planets within the HZ of compact planetary systems can retain such moons (Kane 2017). The two main considerations for the region in which a moon can be maintained in orbit around a planet are the Hill sphere radius (the maximum distance from a planet where its gravitational influence dominates over that of the star) and the Roche limit (the minimum distance from a planet that a moon can avoid being torn apart by the planet's gravity). The Hill radius, R_H, is given by

$$R_H = a_p \left(\frac{M_p}{3M_\star} \right)^{\frac{1}{3}} \tag{5.15}$$

where a_p is the planetary semimajor axis, M_p is the planetary mass, and $M_\star$ is the stellar mass. The Roche limit, R_R, is given by

$$R_R \simeq 2.44 R_p \left(\frac{\rho_p}{\rho_m} \right)^{\frac{1}{3}} \tag{5.16}$$

where R_p is the planetary radius, ρ_p is the planetary density, and ρ_m is the moon density. For example, the Hill radius of the Earth is 0.01 AU and the Roche limit is 1.26×10^{-4} AU, a factor of ~78.5 difference. For an Earth-analog in a compact system with a semimajor axis of 0.05 AU, the Hill radius shrinks to 6×10^{-4} AU, which results in a factor of ~4.5 difference with the Roche limit. Thus it is unlikely that the presence of moons for planets in compact systems plays a role in their possible habitability.

For the second scenario, in which a potentially habitable moon orbits a giant planet, there are various sources of energy that contribute to the total energy budget (Hinkel & Kane 2013; Heller 2012; Heller & Barnes 2013; Heller et al. 2014). In addition to the flux received from the host star (Equation (4.1)), the giant planet provides reflected light from the host star, and IR flux from thermal emission. Shown in Figure 5.8 is the full flux phase profile for an example that considers an Earth-sized moon orbiting a Jupiter-sized planet located in a 1 AU circular orbit around a solar analog (Hinkel & Kane 2013). The horizontal axis represents one complete orbit of the giant planet. The three flux components of stellar, reflective,

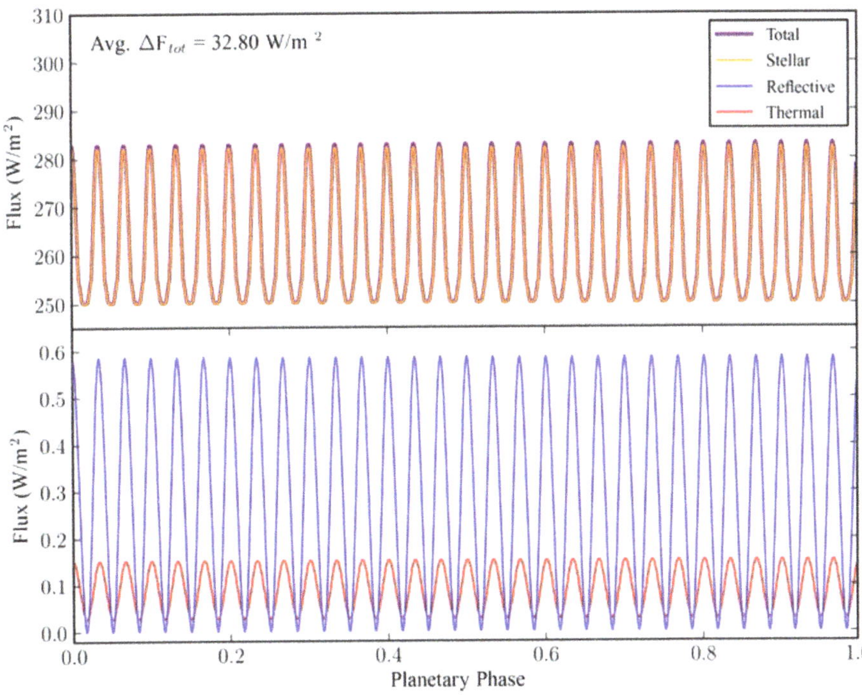

Figure 5.8. Flux phase profile of an example Earth-sized moon orbiting at 0.01 AU around a Jupiter-sized planet, at 1 AU from the Sun, for one complete orbit of the planet. The contributions to the received energy by the Moon include fluxes from the star, and both reflected light and thermal emission from the giant planet. The top panel shows the combined fluxes from all three components, while the bottom panel shows only those contributions from the giant planet. The average total flux variation on the Moon is given in the top left corner. Credit: Reproduced from Hinkel & Kane (2013). © 2020. The American Astronomical Society. All rights reserved.

and thermal are shown in the top panel, and the bottom panel shows the specific contributions of the reflective and thermal components. In this case, the incident flux on the Moon is overwhelmingly dominated by the stellar flux, although giant planets will have strong thermal emission early in their evolutionary history (Fortney & Nettelmann 2010).

A significant component that is not included in Figure 5.8 is the tidal energy provided to the Moon via the gravitational forces exerted by the giant planet (Heller & Barnes 2013). Perhaps the best observed example of such tidal forces are the dramatic effect that Jupiter's gravity has on its moon Io, inducing significant volcanic outgassing (Peale et al. 1979; Lainey et al. 2009). Note that the orbital resonances among the Galilean moons serve to excite the eccentricity of Io and thus sustain the high level of tidal forces acting within it (Greenberg 1987). Figure 5.9 shows the average total energy received at the top of the atmosphere for a 0.5 Earth-mass moon that is orbiting a 5 Jupiter-mass planet (Dobos et al. 2017). The planet is assumed to lie in a 0.1 eccentricity orbit at 1 AU around a Sun-like star. The total energy budget includes contributions from stellar radiation, the planetary reflected and thermal radiation, and tidal heating, where the tidal component was calculated using a constant-time-lag (CTL) model. For comparison, the orbital periods of

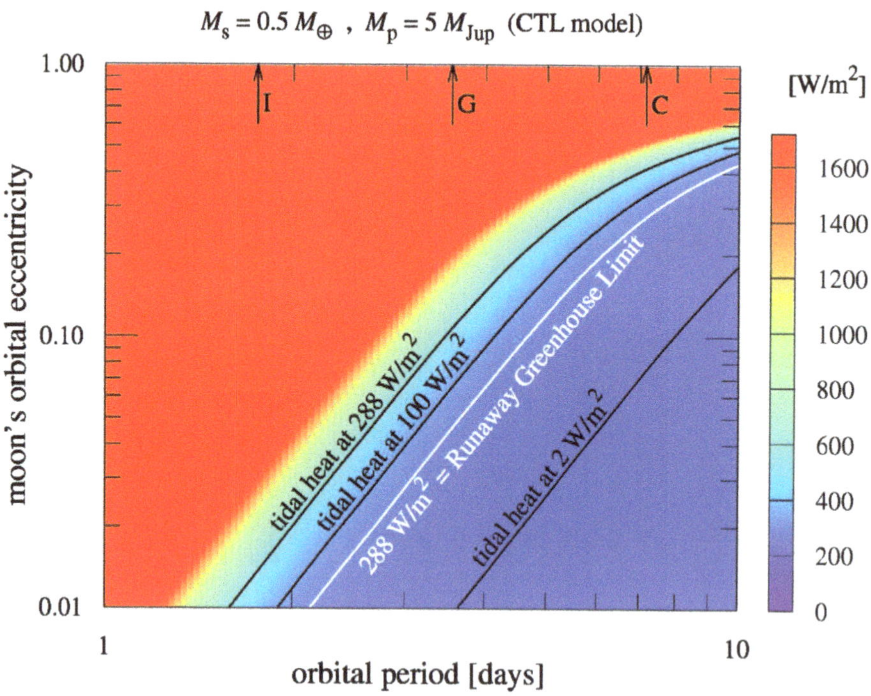

Figure 5.9. Energy flux at the top of the atmosphere for a 0.5 Earth-mass moon orbiting a 5 Jupiter-mass planet in a 1 AU, 0.1 eccentricity orbit around a Sun-like star. The white curve indicates the runaway greenhouse limit considering all energy sources, while the black curves represent the tidal heating energy component alone. The arrows at the top of the figure indicate the orbital periods of Io (I), Ganymede (G), and Callisto (C). Credit: Dobos et al. (2017), reproduced with permission © ESO.

Io (I), Ganymede (G), and Callisto (C) are also indicated via arrows at the top of the diagram. The black curves show the tidal heating energy contribution alone, and the white curve shows the combined energy of tidal heating and all other energy sources. The significance of the 288 $\mathrm{W\,m^{-2}}$ is the limit whereby a runaway greenhouse scenario may be triggered for an Earth-sized moon (Kopparapu et al. 2013; Kodama et al. 2018). Indeed, tidal forces have the potential to tip the energy balance of terrestrial planets in eccentric orbits or terrestrial moons around giant planets into the runaway greenhouse regime (Barnes et al. 2013; Heller & Barnes 2015).

References

Agnew, M. T., Maddison, S. T., Horner, J., & Kane, S. R. 2019, MNRAS, 485, 4703
Armstrong, J. C., Barnes, R., Domagal-Goldman, S., et al. 2014, AsBio, 14, 277
Auclair-Desrotour, P., & Heng, K. 2020, A&A, 638, A77
Barnes, R. 2017, CeMDA, 129, 509
Barnes, R. K., Mullins, C., Goldblatt, V. S., & Meadows, J. F. 2013, AsBio, 13, 225
Barnes, R., & Quinn, T. 2004, ApJ, 611, 494
Batygin, K., & Laughlin, G. 2015, PNAS, 112, 4214
Berger, A. 1980, VA, 24, 103
Bolmont, E., Libert, A.-S., Leconte, J., & Selsis, F. 2016, A&A, 591, A106
Bonfils, X., Delfosse, X., Udry, S., et al. 2013, A&A, 549, A109
Bonomo, A. S., Desidera, S., Benatti, S., et al. 2017, A&A, 602, A107
Carrera, D., Raymond, S. N., & Davies, M. B. 2019, A&A, 629, L7
Chambers, J. E. 1999, MNRAS, 304, 793
Chatterjee, S., Ford, E. B., Matsumura, S., & Rasio, F. A. 2008, ApJ, 686, 580
Clement, M. S., Deienno, R., Kaib, N. A., et al. 2021a, Icar, 367, 114556
Clement, M. S., Raymond, S. N., Kaib, N. A., Deienno, R., & Chambers, J. E. 2021b, Icar, 355, 114122
Colose, C. M., Del Genio, A. D., & Way, M. J. 2019, ApJ, 884, 138
Cumming, A., Butler, R. P., Marcy, G. W., et al. 2008, PASP, 120, 531
Deitrick, R., Barnes, R., Bitz, C., et al. 2018a, AJ, 155, 266
Deitrick, R., Barnes, T. R., Quinn, J., et al. 2018b, AJ, 155, 60
Dobos, V., Heller, R., & Turner, E. L. 2017, A&A, 601, A91
Dressing, C. D., Spiegel, D. S., Scharf, C. A., Menou, K., & Raymond, S. N. 2010, ApJ, 721, 1295
Duncan, M. J., Levison, H. F., & Lee, M. H. 1998, AJ, 116, 2067
Endl, M., Cochran, W. D., Kürster, M., et al. 2006, ApJ, 649, 436
Ford, E. B. 2014, PNAS, 111, 12616
Ford, E. B., & Rasio, F. A. 2008, ApJ, 686, 621
Forgan, D. 2016, MNRAS, 463, 2768
Forgan, D., & Yotov, V. 2014, MNRAS, 441, 3513
Fortney, J. J., & Nettelmann, N. 2010, SSRv, 152, 423
Fulton, B. J., Rosenthal, L. J., Hirsch, L. A., et al. 2021, ApJS, 255, 14
Georgakarakos, N., Eggl, S., & Dobbs-Dixon, I. 2018, ApJ, 856, 155
Gillon, M., Triaud, A. H. M. J., Demory, B.-O., et al. 2017, Natur, 542, 456
Greenberg, R. 1987, Icar, 70, 334
Guzewich, S. D., Lustig-Yaeger, J., Davis, C. E., et al. 2020, ApJ, 893, 140

Hammond, M., & Lewis, N. T. 2021, PNAS, 118, 2022705118
Hatzes, A. P. 2016, SSRv, 205, 267
He, M. Y., Ford, E. B., & Ragozzine, D. 2019, MNRAS, 490, 4575
Heller, R. 2012, A&A, 545, L8
Heller, R., & Barnes, R. 2015, IJAsB, 14, 335
Heller, R., Leconte, J., & Barnes, R. 2011, A&A, 528, A27
Heller, R., & Barnes, R. 2013, AsBio, 13, 18
Heller, R., Williams, D., Kipping, D., et al. 2014, AsBio, 14, 798
Hinkel, N. R., & Kane, S. R. 2013, ApJ, 774, 27
Horner, J., & Jones, B. W. 2008, IJAsB, 7, 251
Horner, J., Vervoort, P., Kane, S. R., et al. 2020, AJ, 159, 10
Jansen, T., Scharf, C., Way, M., & Del Genio, A. 2019, ApJ, 875, 79
Johnson, J. A., Aller, K. M., Howard, A. W., & Crepp, J. R. 2010, PASP, 122, 905
Jurić, M., & Tremaine, S. 2008, ApJ, 686, 603
Kane, S. R. 2015, ApJ, 814, L9
Kane, S. R. 2017, ApJ, 839, L19
Kane, S. R., Ciardi, D. R., Gelino, D. M., & von Braun, K. 2012, MNRAS, 425, 757
Kane, S. R., & Gelino, D. M. 2012, AsBio, 12, 940
Kane, S. R., & Raymond, S. N. 2014, ApJ, 784, 104
Kane, S. R., & Torres, S. M. 2017, AJ, 154, 204
Kane, S. R., Wittenmyer, R. A., Hinkel, N. R., et al. 2016, ApJ, 821, 65
Kane, S. R., Arney, G. N., Byrne, P. K., et al. 2021, JGRE, 126, e06643
Kane, S. R., & Blunt, S. 2019, AJ, 158, 209
Kane, S. R., Li, Z., Wolf, E. T., Ostberg, C., & Hill, M. L. 2021, AJ, 161, 31
Kane, S. R., Turnbull, M. C., Fulton, B. J., et al. 2020, AJ, 160, 81
Kinoshita, H., Yoshida, H., & Nakai, H. 1991, CeMDA, 50, 59
Kipping, D. M., Schmitt, A. R., Huang, X., et al. 2015, ApJ, 813, 14
Kite, E. S., Gaidos, E., & Manga, M. 2011, ApJ, 743, 41
Kodama, T., Nitta, A., Genda, H., et al. 2018, JGRE, 123, 559
Koll, D. D. B., & Abbot, D. S. 2016, ApJ, 825, 99
Kopparapu, R. K., & Barnes, R. 2010, ApJ, 716, 1336
Kopparapu, R. K., Ramirez, R., Kasting, J. F., et al. 2013, ApJ, 765, 131
Kopparapu, R. K., Raymond, S. N., & Barnes, R. 2009, ApJ, 695, L181
Lainey, V., Arlot, J.-E., Karatekin, Ö., & Hoolst, T. V. 2009, Natur, 459, 957
Laskar, J. 1994, A&A, 287, L9
Laskar, J., Joutel, F., & Robutel, P. 1993, Natur, 361, 615
Laskar, J., & Robutel, P. 1993, Natur, 361, 608
Laskar, J., Robutel, P., Joutel, F., et al. 2004, A&A, 428, 261
Laughlin, G., Deming, D., Langton, J., et al. 2009, Natur, 457, 562
Li, G., & Batygin, K. 2014, ApJ, 790, 69
Linsenmeier, M., Pascale, S., & Lucarini, V. 2015, P&SS, 105, 43
Lisiecki, L. E. 2014, PalOc, 29, 71
Lisiecki, L. E., & Raymo, M. E. 2005, PalOc, 20, PA1003
Lissauer, J. J., Barnes, J. W., & Chambers, J. E. 2012, Icar, 217, 77
Lissauer, J. J., Quintana, E. V., Rivera, E. J., & Duncan, M. J. 2001, Icar, 154, 449
Luger, R., Sestovic, M., Kruse, E., et al. 2017, NatAs, 1, 0129

Lunine, J. I. 2001, PNAS, 98, 809
Matsumura, S., Ida, S., & Nagasawa, M. 2013, ApJ, 767, 129
Méndez, A., & Rivera-Valentín, E. G. 2017, ApJ, 837, L1
Morbidelli, A., & Raymond, S. N. 2016, JGRE, 121, 1962
Obertas, A., Laerhoven, C. V., & Tamayo, D. 2017, Icar, 293, 52
Olsen, P. E., & Kent, D. V. 1999, RSPTA, 357, 1761
Palubski, I. Z., Shields, A. L., & Deitrick, R. 2020, ApJ, 890, 30
Peale, S. J., Cassen, P., & Reynolds, R. T. 1979, Sci, 203, 892
Quinn, T. R., Tremaine, S., & Duncan, M. 1991, AJ, 101, 2287
Rasio, F. A., & Ford, E. B. 1996, Sci, 274, 954
Rein, H., & Liu, S. F. 2012, A&A, 537, A128
Rein, H., & Tamayo, D. 2015, MNRAS, 452, 376
Reynolds, R. T., McKay, C. P., & Kasting, J. F. 2015, AdSpR, 7, 125
Rosengren, A. J., & Scheeres, D. J. 2014, CeMDA, 118, 197
Saha, P., & Tremaine, S. 1992, AJ, 104, 1633
Sergeev, D. E., Lambert, F. H., Mayne, N. J., et al. 2020, ApJ, 894, 84
Shen, Y., & Turner, E. L. 2008, ApJ, 685, 553
Shields, A. L., Ballard, S., & Johnson, J. A. 2016, PhR, 663, 1
Smith, A. W., & Lissauer, J. J. 2009, Icar, 201, 381
Smith, A. W., & Lissauer, J. J. 2010, CeMDA, 107, 487
Spiegel, D. S., Raymond, S. N., Dressing, C. D., Scharf, C. A., & Mitchell, J. L. 2010, ApJ, 721, 1308
Tomasella, L., Marzari, F., & Vanzani, V. 1996, P&SS, 44, 427
Tsiganis, K., Gomes, R., Morbidelli, A., & Levison, H. F. 2005, Natur, 435, 459
Van Eylen, V., & Albrecht, S. 2015, ApJ, 808, 126
Wang, S., Xuesong,, Wright, J. T., et al. 2012, ApJ, 761, 46
Way, M. J., & Georgakarakos, N. 2017, ApJ, 835, L1
Williams, D. M. 2003, AmJPh, 71, 1198
Williams, D. M., & Kasting, J. F. 1997, Icar, 129, 254
Williams, D. M., Kasting, J. F., & Wade, R. A. 1997, Natur, 385, 234
Williams, D. M., & Pollard, D. 2002, IJAsB, 1, 61
Williams, D. M., & Pollard, D. 2003, IJAsB, 2, 1
Winn, J. N., & Fabrycky, D. C. 2015, ARA&A, 53, 409
Wittenmyer, R. A., Butler, R. P., Tinney, C. G., et al. 2016, ApJ, 819, 28
Wittenmyer, R. A., Butler, R. P., Horner, J., et al. 2020, MNRAS, 491, 5248
Wordsworth, R. 2015, ApJ, 806, 180
Zechmeister, M., Kürster, M., Endl, M., et al. 2013, A&A, 552, A78

Stephen R Kane

Chapter 6

The Habitable Zone

6.1 Controversial Terminology

The habitable zone (HZ) is one of the most, if not the most, controversial terms that currently exists within the field of astrobiology. Much of the controversy relates to confusion and/or disagreement as to what the term actually means. For example, the phrase "habitable zone" implies "the zone in which bodies are habitable," but that face-value interpretation is incorrect. Because of this confusion, some have suggested that the terminology be changed to something like "liquid water zone," but even that presents a misleading idea of what information the region is intended to provide. Thus, it is important that the motivation and purpose of the HZ be clearly formulated so that it is used in both a valid and effective manner.

The concept of the HZ has existed within the scientific literature for many decades before the first exoplanets were discovered (Huang 1959). Initially the concept was vague with regards to the criteria for the HZ and the boundaries that encompassed it. This was partly because of the evolving nature of astrobiology as a scientific discipline, but was also constrained due to the relative lack of computing power needed to fully realize a robust definition. Assessing a requirement of surface liquid water involves calculations of atmospheric radiative transfer, and solving the response of the surface and atmosphere to changing climate conditions. A significant advance was achieved in the early 1990s when the use of one-dimensional climate models became more accessible and were applied to the HZ problem by James Kasting and collaborators (Kasting 1988, 1991; Kasting et al. 1993). This model assumed an Earth-like planet with a $CO_2/H_2O/N_2$ atmosphere and calculated the loss of surface water as a function of the intensity and spectral energy distribution of the incident flux. The inner boundary was then defined by a water dominated atmosphere, eventually losing water via photolysis and hydrogen escape, and the outer boundary by an increase in the albedo due to sufficient CO_2 cloud formation, which dominates the CO_2 greenhouse warming. These boundaries thus quantified those locations where runaway greenhouse and snowball scenarios would occur (Selsis et al. 2007).

doi:10.1088/2514-3433/ac2aa1ch6

However, clearly not all terrestrial planets can be assumed to be Earth-like, and in fact may differ in many respects, such as size, mass, atmospheric composition, age, interior, magnetic field, etc. There have been numerous published modifications to the HZ, including consideration of relatively dry planets (Abe et al. 2011), the effects of planetary rotation (Yang et al. 2014; Leconte et al. 2015; Kopparapu et al. 2016), stellar evolution (Ramirez & Kaltenegger 2014), and stellar multiplicity (Haghighipour & Kaltenegger 2013; Kane & Hinkel 2013; Cukier et al. 2019). So what really is the utility of the HZ given the vast ocean of parameter space, some of which is assumed, and much of which cannot be presently measured (see Table 2.1)? To answer this, we must carefully examine the definition of the HZ. *The HZ is defined as the region around a star where water could exist in a liquid state on the surface of a planet if it has sufficient atmospheric pressure.* There are several key components of this definition that should be clarified.

1. "Region around a star." As noted above, and discussed in detail in Chapter 4, although there are many planetary properties that influence surface conditions, the primary observable and energy source is the parent star. Thus, the power output of the star is the basis of the HZ calculations.
2. "Water could exist in a liquid state on the surface." As discussed in Section 4.1, the use of surface liquid water as a requirement creates quantifiable boundary conditions under which this requirement is met. Furthermore, Section 1.2 outlines some of the reasons that liquid water is specifically chosen, including its prevalence, relatively high temperature, and of course pivotal role in the evolution of Earth-based life.
3. "Sufficient atmospheric pressure." Beyond the assumption that the planet is similar to Earth, the requirement of an atmosphere with surface liquid water allows the prediction of the surface energy balance and measurable atmospheric signatures. The latter of these is crucial in establishing an observational program that can eventually validate the model predictions. This remote sensing requirement is why sub-surface oceans are not included in the HZ definition.

As described in Section 1.3, the number of known exoplanets is rapidly increasing, and terrestrial planets far outnumber their giant planet counterparts. Hence, the purpose of the HZ is as a target selection tool to guide the limited resources that will be able to observe a subset of the known planets. Planetary habitability is a complex subject, and there are numerous reasons why a planet outside the HZ may be habitable, and why a planet inside the HZ may be desolate. However, the HZ provides us with a "first guess" where we may optimize our chances of detecting favorable conditions for life as we know it. Eventually, as our sample size increases and our measured understanding of planetary habitability grows, the need for the HZ concept will gradually disappear.

6.2 Defining the Search Space

To describe the boundaries of the HZ, we will adopt the formalism used by Ravi Kopparapu and collaborators in their updated one-dimensional climate model, which remain in common use (Kopparapu et al. 2013, 2014). The calculations of this model use a radiative transfer analysis that relies on an accurate understanding of the absorption and emission properties of molecular species within the atmosphere. These properties are provided by the high-resolution transmission molecular absorption database (HITRAN; Rothman et al. 2009, 2013) and the high-temperature molecular spectroscopic database (HITEMP; Rothman et al. 2010). HITRAN/HITEMP is a compilation of spectroscopic parameters used to predict and simulate transmission and emission within the atmosphere. The model includes Rayleigh scattering by water vapor and has been validated with the Spectral Mapping Atmospheric Radiative Transfer (SMART) model (Meadows & Crisp 1996; Crisp 1997).

The calculation of the inner HZ boundaries uses the effective solar flux, $S_{\rm eff}$, which is the ratio of the net incoming solar flux, $F_{\rm SOL}$, and net outgoing IR flux, $F_{\rm IR}$, at the top of the atmosphere, and is thus related to the atmospheric albedo. By inserting the solar luminosity ($L_\odot \sim 3.828 \times 10^{26}$ W m^{-2}) and Earth's semimajor axis ($a_\oplus \sim 1.495\,978\,7 \times 10^{11}$ m) into Equation (4.1), we find the flux received at the top of Earth's atmosphere, referred to as the solar constant, to be ~1361 W m^{-2}. This incident flux reaches the area of a circle, $\pi R_\oplus^2$ but, considering the area of the planet, $4\pi R_\oplus^2$, the average incident flux received is a factor of 4 smaller than the solar constant, or ~340 W m^{-2}. The top of the atmosphere flux is further refined by considering a Bond albedo of $A \sim 0.3$, resulting in a net incoming solar flux of ~231 W m^{-2}. This simple calculation places the values for $F_{\rm SOL}$ and $F_{\rm IR}$ shown in the top-left panel of Figure 6.1 in context, where the fluxes are calculated to preserve a particular surface temperature in the range 200–2200 K. The $F_{\rm IR}$ initially increases with surface temperature, then plateaus as the atmosphere becomes opaque to IR radiation, then increases again as the lower atmosphere and surface begin to radiate in the visible and NIR. $F_{\rm SOL}$ also initially increases due to NIR absorption by water vapor, then decreases to a constant value at higher temperatures as a result of Rayleigh scattering. As expected, the albedo (shown in the top-right panel) mirrors $F_{\rm SOL}$, with a minimum value at a temperature of ~400 K. The variation in $S_{\rm eff}$ is shown in the bottom-left panel, from which the HZ boundaries are calculated. For example, the "moist greenhouse," at a surface temperature of 340 K and $S_{\rm eff} = 1.015$, occurs when the stratospheric water vapor content increases by an order of magnitude. This water vapor increase is also apparent in the bottom-right panel when comparing H_2O volume mixing ratio between surface temperatures of 320 K and 340 K. The runaway greenhouse limit, where $S_{\rm eff} = 1.06$, occurs when the oceans evaporate entirely. The outer HZ boundary is calculated by considering the atmospheric CO_2 partial pressure, $p{\rm CO}_2$, which is assumed to fluctuate in the atmosphere in response to temperature variations as a result of the carbonate-silicate cycle. Applying this to the model above reveals that a maximum greenhouse limit occurs when $S_{\rm eff} = 0.325$, indicating that further decreases in $F_{\rm SOL}$ are unable to sustain surface temperatures that would prevent freezing of surface liquid water. Note that these calculations for the outer boundary do not include radiative warming by CO_2 clouds, and thus should be considered conservative.

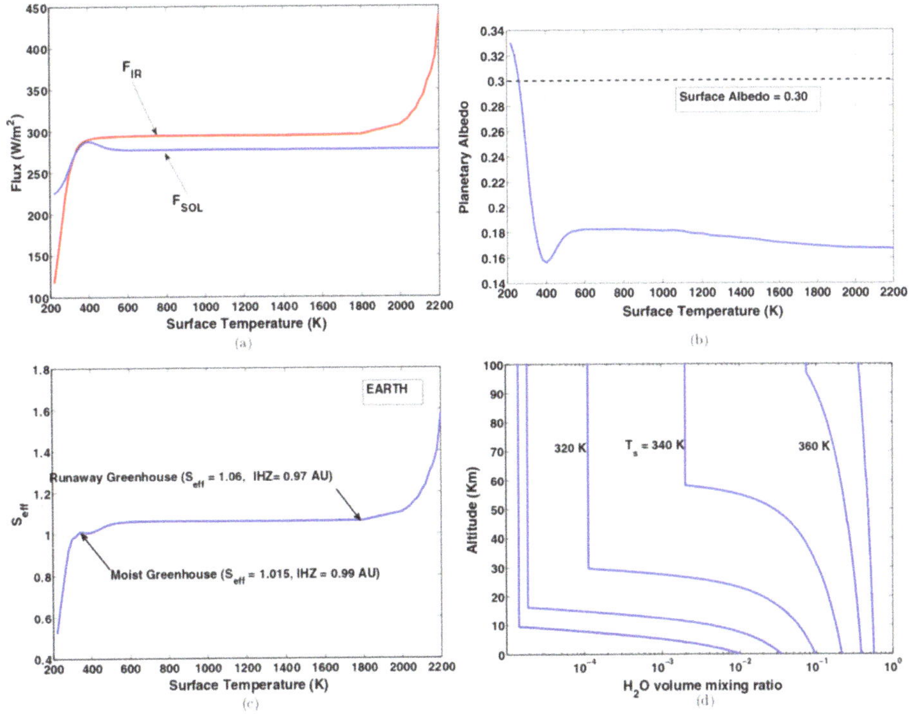

Figure 6.1. Inner HZ boundary (IHZ) calculations as a function of surface temperature. Top left: net outgoing IR flux and net incident solar flux. Top right: planetary albedo. Bottom left: effective solar flux, $S_{\mathrm{eff}} = F_{\mathrm{IR}}/F_{\mathrm{SOL}}$. Bottom right: water vapor profile. Credit: Reproduced from Kopparapu et al. (2013). © 2013. The American Astronomical Society. All rights reserved.

The model described above results in four major HZ boundaries; two inner and two outer. For the inner edge, these are the runaway greenhouse limit and the "recent Venus" limit, based on the empirical observation that the Venusian surface may have had liquid water as recently as a billion years ago. For the outer edge, the boundaries consist of the maximum greenhouse limit and the "early Mars" limit, based on the observation that Mars may have had surface liquid water 3.8 Gyr ago when solar luminosity was ~25% lower. The runaway greenhouse and maximum greenhouse boundaries encapsulate the Conservative Habitable Zone (CHZ), while the recent Venus and early Mars boundaries encapsulate the Optimistic Habitable Zone (OHZ; Kane et al. 2016). The distance of a particular HZ boundaries from the host star, d, is specified by:

$$d_{\mathrm{HZ}} = \left(\frac{L_\star/L_\odot}{S_{\mathrm{eff}}}\right)^{0.5} \mathrm{AU} \tag{6.1}$$

where the effective stellar flux, S_{eff}, is given by:

$$S_{\mathrm{eff}} = S_{\mathrm{eff}\odot} + c_1 T_\star + c_2 T_\star^2 + c_3 T_\star^3 + c_4 T_\star^4 \tag{6.2}$$

and $T_\star = T_{\mathrm{eff}} - 5780$ K. The polynomial coefficients, c_1 through c_4, for the CHZ and OHZ boundaries depend on updates to the HITEMP database, and the most recent versions are available online (Kopparapu et al. 2014). As formulated in Equations (6.1) and (6.2), and as described in Section 6.1, the HZ boundaries make assumptions regarding the planetary properties and depend only on the stellar luminosity and effective temperature. Note also that the above calculations are valid for stellar temperatures in the range 2600 K $\leqslant T_{\mathrm{eff}} \leqslant$ 7200 K.

The components of the CHZ and OHZ boundaries are shown in the top panel of Figure 6.2, as a function of stellar effective temperature and incident flux on

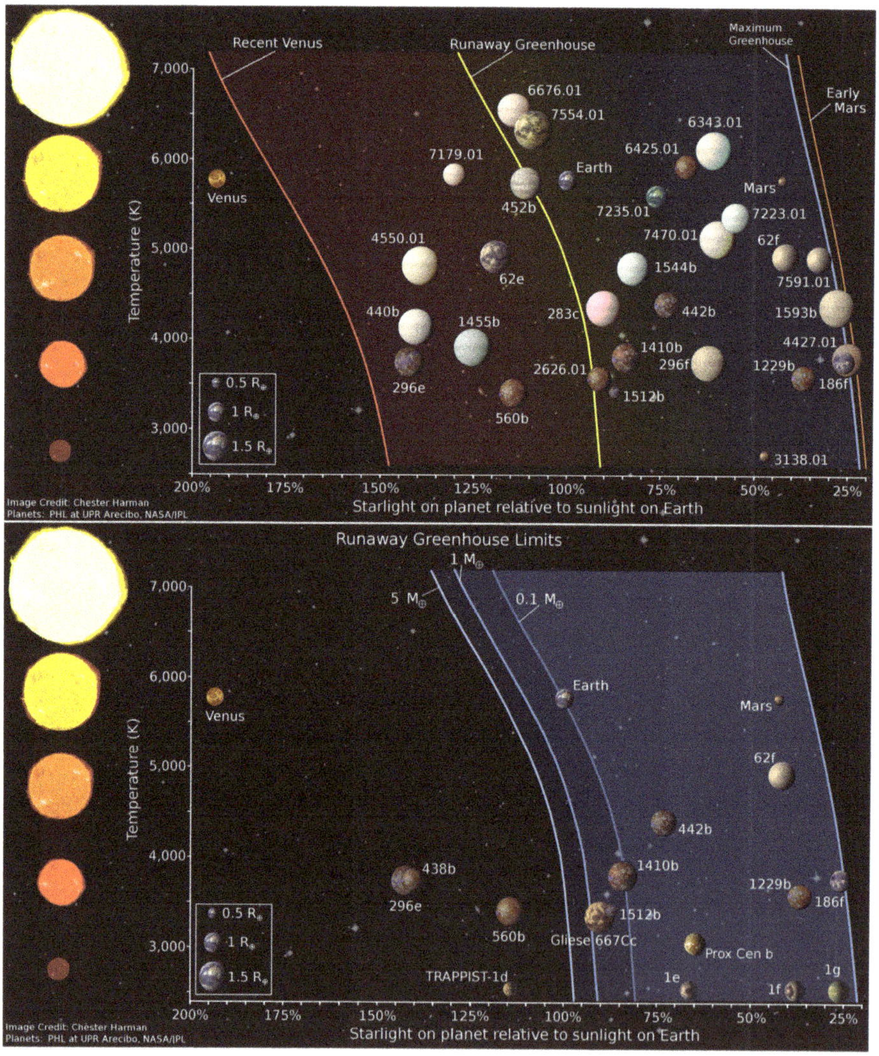

Figure 6.2. Top: stellar effective temperature as a function of incident flux for planetary candidates discovered using data from the Kepler mission (Kane et al. 2016). The relative sizes of the planets are shown, along with their Kepler designations, and are overplotted on the conservative and optimistic HZ regions. Bottom: similar to the top panel, but showing the effect of planetary mass on the runaway greenhouse limit (Kopparapu et al. 2014). Credit: Reproduced with permission from Chester Harman.

the planet relative to Earth (Kane et al. 2016). Discoveries from the Kepler mission that lie within the HZ are overlaid and shown with their relative sizes, along with their Kepler designations. The location and sizes of Venus, Earth, and Mars are included for comparison. The top panel of Figure 6.2 also demonstrates the dramatic effect that the recent Venus boundary has on extending the extent of the HZ inward toward the host star. This allows the consideration of substantially more exoplanets, particularly as the common detection methods of radial velocities and transits are highly biased toward shorter orbital periods (see Section 2.5 and Chapter 8).

A modification to the above HZ boundaries may be calculated by scaling the planetary mass and adjusting the atmospheric pressure accordingly (Kopparapu et al. 2014). Assuming that the fraction of volatiles outgassed during or after accretion is the same for all planetary masses, the relationship between surface pressure, mass, and radius, may be expressed as:

$$\frac{P_s}{P_\oplus} = \left(\frac{M_p}{M_\oplus}\right)^2 \left(\frac{R_\oplus}{R_p}\right)^4 \tag{6.3}$$

where $P_\oplus$ is the surface pressure on Earth. By adopting a suitable mass–radius relationship for terrestrial exoplanets (see Section 4.3), this can be used to provide a relationship between planetary radius and surface pressure. The surface pressure will thus increase with planet size and, according to Equation (4.11), an increase in surface gravity results in a decrease in atmospheric scale height. The net result is an increase in H_2O column depth for low-mass planets, which is equivalent to an increase in $S_{\rm eff}$ for high-mass planets. Therefore, the inner boundary of the CHZ (runaway greenhouse limit) moves closer to the star for high-mass planets, as shown in the bottom panel of Figure 6.2. Notice that the outer CHZ boundary (maximum greenhouse limit) remains relatively unaffected.

Figure 6.3 shows top-down views of four known exoplanetary systems with orbits within the HZ, where the size along a panel side (scale) in the figure is indicated in the top-right corner of each panel. In each case, the host star is located at the center of the dotted cross-hairs and the extent of the HZ is shown in green, where light green is the conservative HZ and dark green is the optimistic extension to the HZ. These data were extracted from the Habitable Zone Gallery[1] (Kane & Gelino 2012). The top-left panel shows the HD 40307 system, that consists of a K dwarf host star and six known super-Earths and mini-Neptune planets (Tuomi et al. 2013). Planet g lies within the HZ on an eccentric orbit ($e = 0.29$) that causes it to move from the CHZ to the inner OHZ, resulting in a variable incident flux (see Section 5.1). The top-right panel shows the HD 45364 system, that consists of a solar-type host star and two known giant planets in eccentric orbits (Correia et al. 2009). Both planetary orbits are within the HZ and are in 3:2 resonance such that they never experience close-encounters, similar to the orbits of Neptune and Pluto (Williams & Benson 1971). The bottom-left panel shows the Kepler-62 system, that consists of a K dwarf host star and five known

[1] http://hzgallery.org/

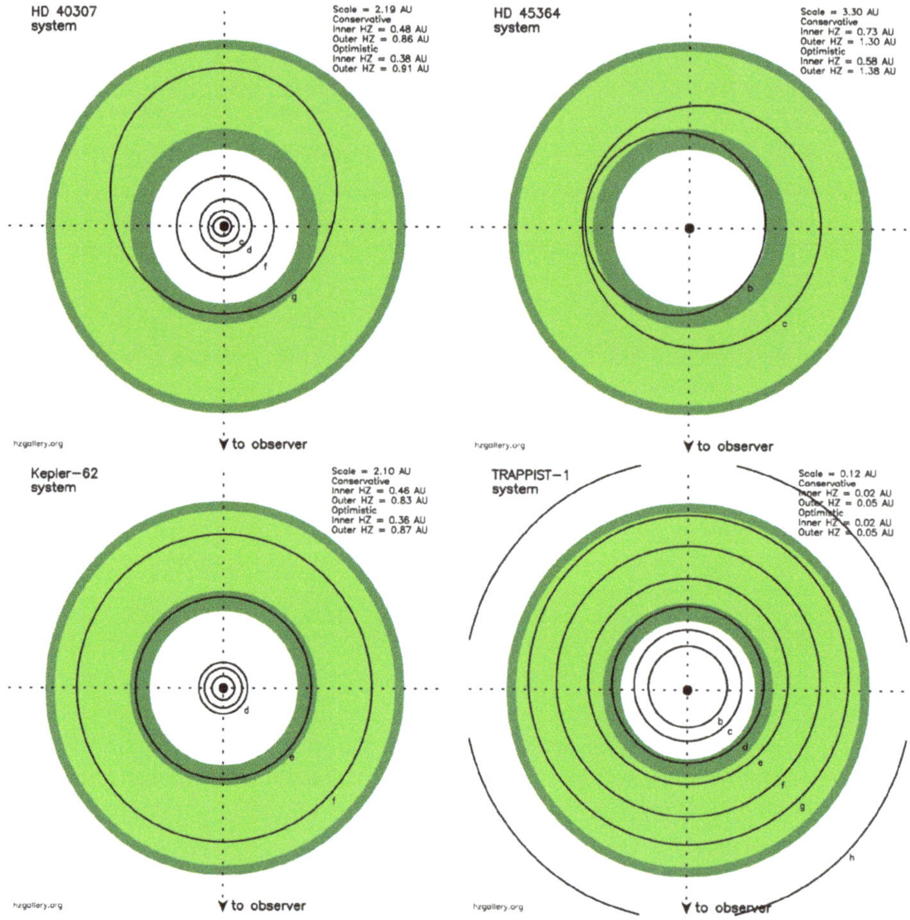

Figure 6.3. Examples of orbits within the HZ for four known exoplanetary systems, showing a top-down view of each system. The orbits are shown as solid lines, and the dotted cross-hairs indicate the location of the host star. The extent of the HZ is shown in green, where light green is the conservative HZ and dark green is the optimistic extension to the HZ.

transiting planets whose sizes range between Mars-size to twice Earth-size. The HZ planets, e and f, are ~50% larger than Earth and lie within the inner OHZ and CHZ, respectively. The bottom-right panel shows the TRAPPIST-1 system, that consists of a cool M dwarf host star and seven known transiting planets whose sizes lie in the range 75%–125% Earth radii (Gillon et al. 2017; Agol et al. 2021). The extraordinary aspect of the TRAPPIST-1 system is that it contains four known terrestrial planets with orbits in the HZ, one of which lies within the inner OHZ.

The four examples in Figure 6.3 are a very small fraction of a rapidly increasing sample of known HZ planets. The rise in HZ planet detections is represented in the histogram shown in Figure 6.4, where each yearly bin has three categories shown in separate colors. The red portion are those planets that lie interior to the HZ and constitute the vast majority of exoplanet detections because of the bias of most

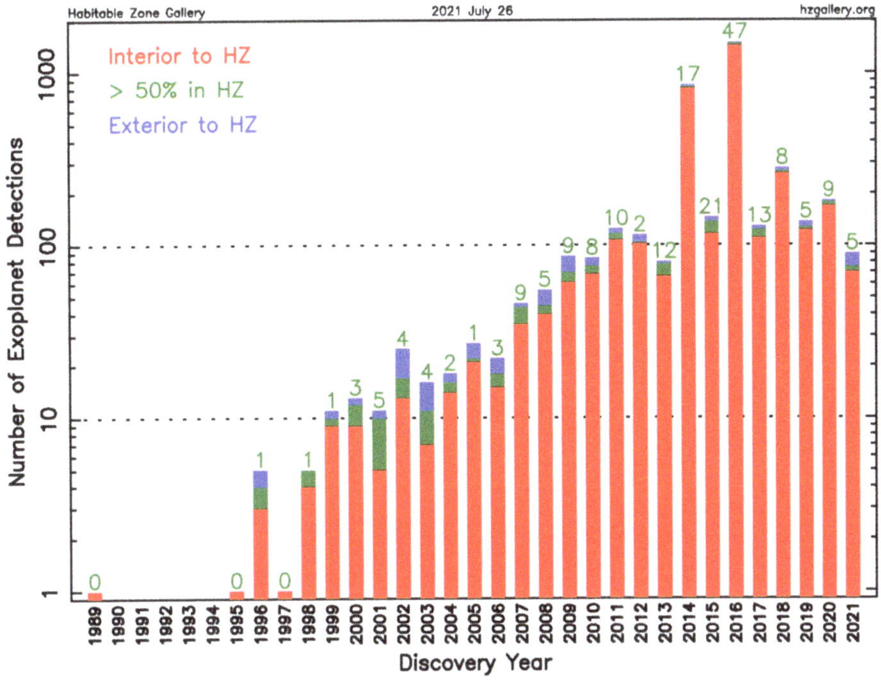

Figure 6.4. Histogram of exoplanet discovery rate, where the discoveries from each year are divided into those interior to the HZ (red), exterior to the HZ (blue), and those that spend more than 50% of their orbit within the HZ (green). Credit: Habitable Zone Gallery. The green number above each bin is the total number of HZ exoplanets discovered in that year.

detection techniques toward short orbital periods. The blue portion are those detected planets exterior to the HZ and are relatively small in number due to the same observational bias reasons. Recall from Kepler's second law (Section 5.1) that planets move faster when close to the star, and slower when farther from the star. For planets whose eccentric orbits causes them to have a partial presence in the HZ, we may calculate the fraction of the orbital period that is inside the HZ. If that orbital period fraction is $\geqslant$50%, then the planet is included in the green portion of each bin. The total number of planets that fall into the HZ category are shown in green above each bin. Note that this histogram includes all planets, regardless of size, and so many of the HZ planets are giant planets. However, these planets may harbor terrestrial moons with habitability prospects (see Section 5.4; Hill et al. 2018).

Since, as demonstrated by Figure 6.4, most of the detected exoplanets are close to their host stars, it is worth considering the effect of tidal locking on the HZ (Barnes 2017). A planet with a 1:1 spin–orbit resonance means that the same hemisphere is in constant daylight, whereas the opposite hemisphere experiences perpetual night. This situation can have significant consequences for the planetary climate evolution (Barnes et al. 2013; Yang et al. 2014) as well as the magnetic field strength (Christensen & Aubert 2006; Driscoll & Barnes 2015) and potential atmospheric erosion (Zhang et al. 2009; Gunell et al. 2018). Figure 6.5 shows HZ boundaries

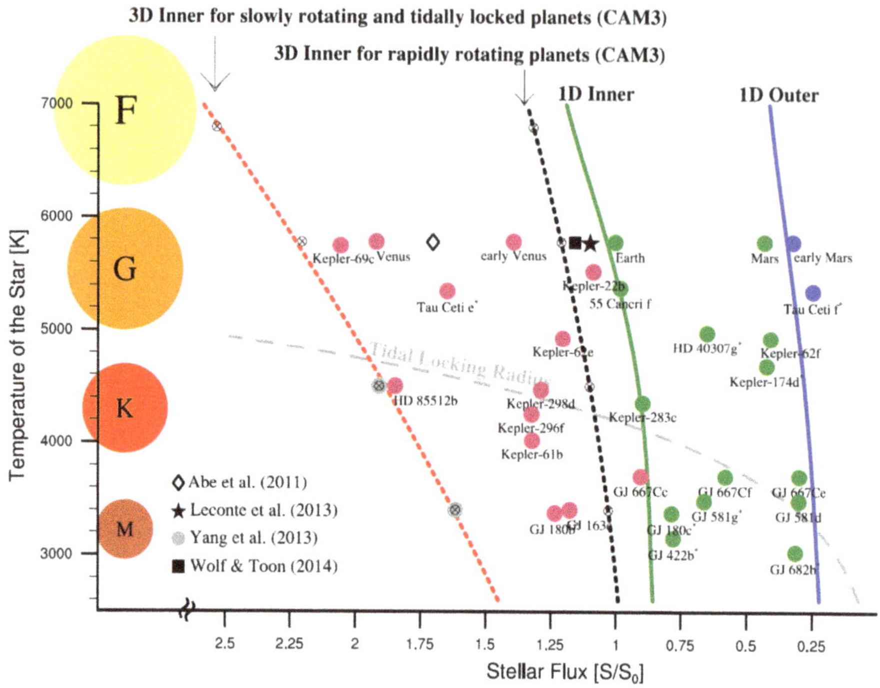

Figure 6.5. The HZ boundaries (solid and dotted lines) and tidal locking distance (gray dashed line) as a function of stellar effective temperature and the stellar flux received by the planet. Solar System planets and numerous known exoplanets are included for comparison. Credit: Reproduced from Yang et al. (2014).

(solid and dotted lines) and tidal locking distance (gray dashed line) as a function of stellar effective temperature and stellar flux received by the planet (Yang et al. 2014). The green and blue solid lines are equivalent to the runaway greenhouse and maximum greenhouse limits shown in the top panel of Figure 6.2. The dotted lines are modified HZ boundaries specifically calculated for various planetary rotational states. Those planets shown below the gray dashed line are considered to be tidally locked. The timescale for tidal locking to occur is given by:

$$\tau_{\text{lock}} = \frac{\omega a^6 I Q}{3 G M_\star^2 k_2 R_p^5} \tag{6.4}$$

where ω is the rotation rate, a is the semimajor axis, I is the moment of inertia, Q is the dissipation function, G is the gravitational constant, $M_\star$ is the mass of the star, k_2 is the Love number, and R_p is the radius of the planet (Gladman et al. 1996). Note that the above calculation neglects the non-negligible interaction that may occur between the planetary atmosphere and the surface (Leconte et al. 2015; Auclair-Desrotour 2017). Even so, it is possible that tidal locking may be more prevalent among exoplanets than expected (Barnes 2017). Furthermore, equations (6.1) and (6.2) show that, for the assumption of an Earth-similar planet, the two main factors in determining the HZ boundaries are the stellar luminosity and effective temperature. However, it is worth considering how well these properties are known to allow accurate HZ calculations (Kane 2014). The flowchart in Figure 3.1 shows that a typical pathway to stellar luminosities is via a combination of apparent brightness and distance (see Section 3.1). Precise distance measurements have often been the limiting factor in HZ boundary determinations, and can lead to an incorrect assessment of the HZ status for terrestrial planets (Kane 2018); a situation that is being greatly improved by data from the Gaia mission (Gaia Collaboration 2021). Effective temperatures are generally more easily accessible and have uncertainties <100 K (Kane 2014). Even so, the uncertainties in stellar parameters should not be ignored in assessing the relative locations of planets within the HZ.

6.3 Evolution Through Time

As a planetary system ages, many facets of the system evolve in ways that impact planetary habitability. The properties of the host star gradually change during its lifetime on the main sequence (Section 3.3); the interior, atmosphere, and outgassing of the planet evolve (Section 4.4); and the orbits of the planets are modified through planet–planet interactions (Section 5.3). For the star, the changes in the luminosity and effective temperature apply directly to Equations (6.1) and (6.2), resulting in changes to the HZ boundaries through time (Underwood et al. 2003; Ramirez & Kaltenegger 2014). This HZ evolution has been studied within the context of stellar masses and chemical compositions (Young et al. 2012; Valle et al. 2014; O'Malley-James et al. 2015; Truitt et al. 2015; Truitt & Young 2017), and also with respect to stellar rotation and magnetic activity (Gallet et al. 2017). This changing aspect of the HZ through time emphasizes the need to consider the full evolutionary history of exoplanets, even if they are found to reside in the HZ at the present epoch.

Shown in Figure 6.6 is a representation of the CHZ and OHZ evolution for an M dwarf (top panel) and G dwarf (bottom panel), showing both pre-main sequence and main sequence phases of stellar evolution. The stellar properties were determined using the MESA Isochrones & Stellar Tracks (MIST) to calculate the evolutionary tracks (Paxton et al. 2011, 2013, 2015; Choi et al. 2016; Dotter 2016; Paxton et al. 2018, 2019). A solar metallicity was assumed, and stellar masses of 0.3 $M_{\odot}$ and 1.0 $M_{\odot}$ were used for the M dwarf and G dwarf, respectively. The boundaries rapidly shift inward as the star transitions to the main sequence, then gradually increase with time. It is worth reminding ourselves here of the dependence of main-sequence lifetimes on stellar mass, as discussed in Sections 3.1 and 3.3. This means that the sub-solar masses of K and M stars will result in a much slower movement of the HZ boundaries than for a G star. Importantly, a terrestrial planet that is near the inner edge of the HZ when the system was young is unlikely to remain within the HZ throughout the main-sequence lifetime, and thus may commence an evolution toward a post-runaway greenhouse state as the star evolves. The dramatic rise in HZ distances shown in the bottom panel at an age of ~10 Gyr is due to the rise in luminosity as the G dwarf star star moves off the main sequence and transitions into the red giant phase. On the other hand, it can be seen from the top panel that an M dwarf maintains relatively stable HZ boundaries for more than 100 Gyr, much longer than the current age of the

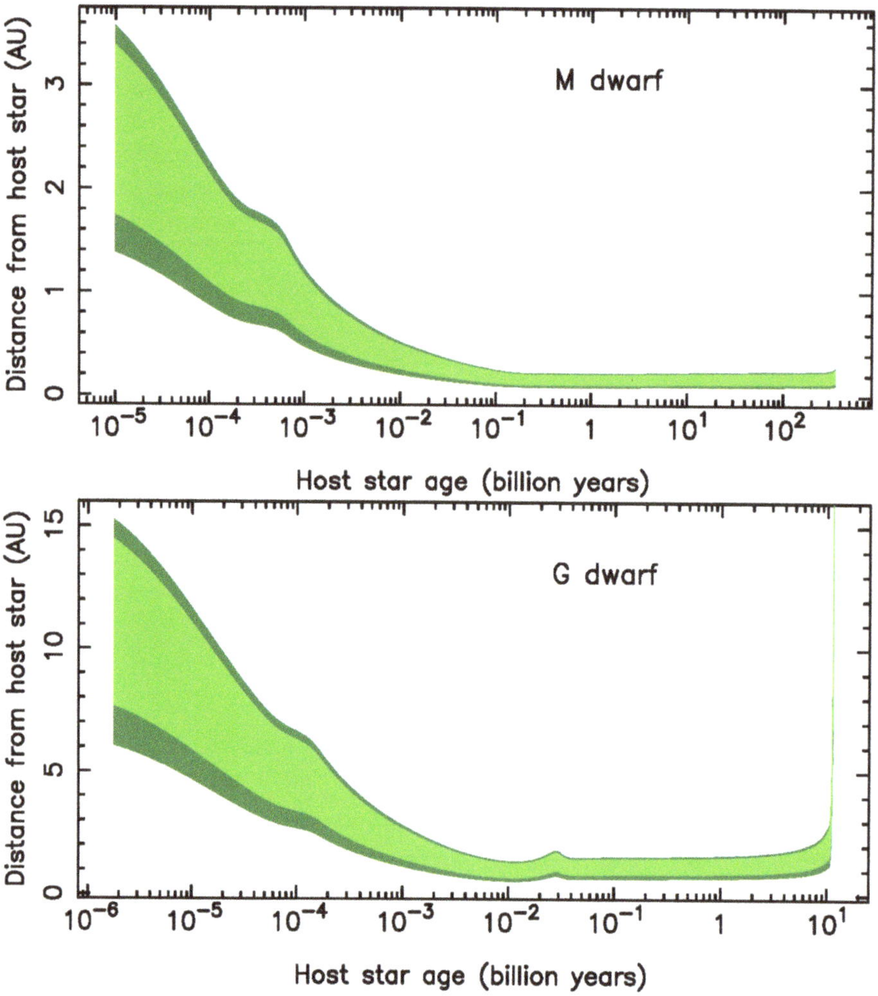

Figure 6.6. Evolution of the HZ before and during the main sequence for an M dwarf (top) and G dwarf (bottom). The extent of the HZ is shown in green, where light green is the conservative HZ and dark green is the optimistic extension to the HZ.

universe. However, since M stars can have prolonged periods of activity that are potentially devastating to planetary atmospheres (see Section 3.2), K dwarf stars may be the perfect balance of spectral energy distribution and longevity.

6.4 Prime Targets

As described in Section 6.1, the primary purpose of the HZ is that of target selection for potentially habitable planets. The required follow-up observations to properly assess the nature of terrestrial planets in the HZ are not trivial and often require access to highly oversubscribed facilities. At the present time, the majority of terrestrial exoplanets have been detected using the transit method, which provides an estimate of the planetary radius. However, the atmospheric scale height (Equation (4.11)), which depends on planet mass, is a key component in the interpretation of planetary spectra (Batalha et al. 2019). The amplitude of the radial velocity signatures, from which planetary mass is often extracted, can be particularly small for HZ terrestrial planets. For example, the motion of Earth around the Sun produces a radial velocity amplitude of $\sim$9 cm s^{-1}, and TRAPPIST-1f (see Figure 6.3), which is less massive than Earth and orbits a much smaller star, has a predicted radial velocity amplitude of $\sim$1 m s^{-1}. This poses an observational challenge for present radial velocity instruments, since most facilities achieve a precision of $\sim$1 m s^{-1} (Fischer et al. 2016), with a handful reaching $\sim$50 cms^{-1} (Robertson et al. 2019; Pepe et al. 2021). Note that the masses of the TRAPPIST-1 planets have been determined through their dynamical interactions with each other, a process called transit timing variations (TTVs), which is particularly efficient for systems with resonant orbits (Agol et al. 2021).

Although the ultimate observational objective for these planets may be considered to be direct imaging (see Chapter 9), an intermediate objective is transmission and emission spectroscopy for those planets that transit their host star (Beichman et al. 2014; Greene et al. 2016; Morley et al. 2017; Batalha et al. 2018). Transmission spectroscopy is based upon the measurement of light from the host star when a planet transits the disk, during which absorption of the light at particular wavelengths reveal atmospheric composition (see Section 4.2). Similarly, emission spectroscopy uses the secondary eclipse (when the planet passes behind the star) to measure the planetary contribution to the total IR flux, and thus provides an estimate of the temperature, depending on assumptions regarding blackbody emission and the effects of the atmosphere. An important target for these spectroscopic observations is the aforementioned TRAPPIST-1 system, whose number

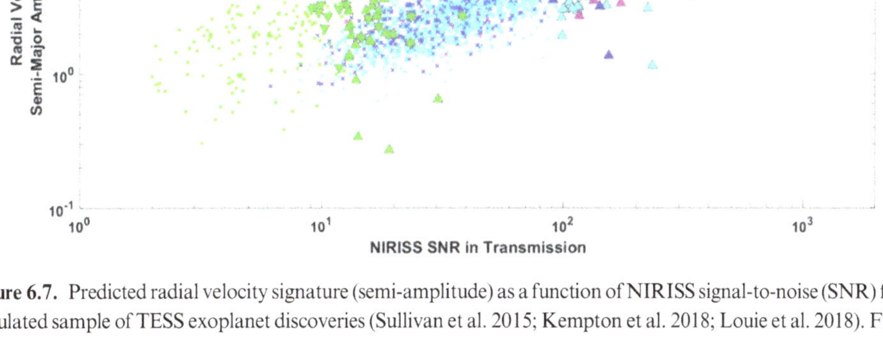

Figure 6.7. Predicted radial velocity signature (semi-amplitude) as a function of NIRISS signal-to-noise (SNR) for a simulated sample of TESS exoplanet discoveries (Sullivan et al. 2015; Kempton et al. 2018; Louie et al. 2018). Filled triangles and upside down triangles are favorable targets for transmission and emission spectroscopy, respectively. The crosses indicate those targets that are disfavored as suitable atmospheric characterization targets. Credit: Reproduced from Kempton et al. (2018). © 2018. The Astronomical Society of the Pacific. All rights reserved.

of HZ planets with relatively short orbital periods provides numerous opportunities for observations to be scheduled (Barstow & Irwin 2016; Fauchez et al. 2019; Lustig-Yaeger et al. 2019). In general, determining suitable targets involves a convolution of the atmospheric scale height, including numerous other planetary and stellar properties, such as the stellar apparent brightness. Shown in Figure 6.7 is the predicted radial velocity signature for a simulated sample of TESS exoplanet discoveries (Sullivan et al. 2015) as a function of the expected signal-to-noise of measurements using the JWST Near Infrared Imager and Slitless Spectrograph (NIRISS) instrument (Louie et al. 2018). The filled triangles represent those planets that are calculated to be optimal transmission spectroscopy targets, and the upside down triangles are those that are best suited as emission spectroscopy targets. The crosses indicate planets for which the signal-to-noise of the observations would be unlikely to yield useful data. For habitability constraints, the radius category of $<1.5\ R_{\oplus}$ is most relevant to our discussion. As expected, these are the most difficult to reach in terms of the expected strength for both their radial velocity and spectroscopic signatures. Even so, these observations will comprise a critical step toward the full characterization of terrestrial planets via direct imaging (Kopparapu et al. 2018) and the necessary testing of the HZ boundaries.

References

Abe, Y., Abe-Ouchi, A., Sleep, N. H., & Zahnle, K. J. 2011, AsBio, 11, 443

Agol, E., Dorn, C., Grimm, S. L., et al. 2021, PSJ, 2, 1

Auclair-Desrotour, P., Laskar, J., Mathis, S., & Correia, A. C. M. 2017, A&A, 603, A108

Barnes, R. 2017, CeMDA, 129, 509

Barnes, R., Mullins, K., Goldblatt, C., et al. 2013, AsBio, 13, 225

Barstow, J. K., & Irwin, P. G. J. 2016, MNRAS, 461, L92

Batalha, N. E., Lewis, N. K., Line, M. R., Valenti, J., & Stevenson, K. 2018, ApJ, 856, L34

Batalha, N. E., Lewis, T., Fortney, J. J., et al. 2019, ApJ, 885, L25

Beichman, C., Benneke, B., Knutson, H., et al. 2014, PASP, 126, 1134

Choi, J., Dotter, A., Conroy, C., et al. 2016, ApJ, 823, 102

Christensen, U. R., & Aubert, J. 2006, GeoJI, 166, 97

Correia, A. C. M., Udry, S., Mayor, M., et al. 2009, A&A, 496, 521

Crisp, D. 1997, GeoRL, 24, 571

Cukier, W., Kopparapu, R. K., Kane, S. R., et al. 2019, PASP, 131, 124402

Dotter, A. 2016, ApJS, 222, 8

Driscoll, P. E., & Barnes, R. 2015, AsBio, 15, 739

Fauchez, T. J., Turbet, M., Villanueva, G. L., et al. 2019, ApJ, 887, 194

Fischer, D. A., Anglada-Escude, G., Arriagada, P., et al. 2016, PASP, 128, 066001

Gaia CollaborationBrown, A. G. A., Vallenari, A., et al. 2021, A&A, 649, A1

Gallet, F., Charbonnel, C., & Amard, L. 2017, A&A, 597, A14

Gillon, M., Triaud, H. M. J., Demory, B.-O., et al. 2017, Natur, 542, 456

Gladman, B., Quinn, D. D., Nicholson, P., & Rand, R. 1996, Icar, 122, 166

Greene, T. P., Line, M. R., Montero, C., et al. 2016, ApJ, 817, 17

Gunell, H., Maggiolo, R., Nilsson, H., et al. 2018, A&A, 614, L3

Haghighipour, N., & Kaltenegger, L. 2013, ApJ, 777, 166

Hill, M. L., Kane, S. R., Seperuelo Duarte, E., et al. 2018, ApJ, 860, 67

Huang, S.-S. 1959, PASP, 71, 421
Kane, S. R. 2014, ApJ, 782, 111
Kane, S. R. 2018, ApJ, 861, L21
Kane, S. R., & Gelino, D. M. 2012, PASP, 124, 323
Kane, S. R., Hill, M. L., Kasting, J. F., et al. 2016, ApJ, 830, 1
Kane, S. R., & Hinkel, N. R. 2013, ApJ, 762, 7
Kane, S. R., Vervoort, P., Horner, J., & Pozuelos, F. J. 2020, PSJ, 1, 42
Kasting, J. F. 1988, Icar, 74, 472
Kasting, J. F. 1991, Icar, 94, 1
Kasting, J. F., Whitmire, D. P., & Reynolds, R. T. 1993, Icar, 101, 108
Kempton, E. M. R., Bean, J. L., Louie, D. R., et al. 2018, PASP, 130, 114401
Kopparapu, R. K., Hebrard, E., Belikov, R., et al. 2018, ApJ, 856, 122
Kopparapu, R. K., Ramirez, R., Kasting, J. F., et al. 2013, ApJ, 765, 131
Kopparapu, R. M., Ramirez, R., Kasting, J. F., et al. 2014, ApJ, 787, L29
Kopparapu, R. K., Wolf, E. T., Haqq-Misra, J., et al. 2016, ApJ, 819, 84
Leconte, J., Wu, H., Menou, K., & Murray, N. 2015, Sci, 347, 632
Louie, D. R., Deming, D., Loic Albert, L., et al. 2018, PASP, 130, 044401
Lustig-Yaeger, J., Meadows, V. S., & Lincowski, A. P. 2019, AJ, 158, 27
Meadows, V. S., & Crisp, D. 1996, JGR, 101, 4595
Morley, C. V., Kreidberg, L., Rustamkulov, Z., Robinson, T., & Fortney, J. J. 2017, ApJ, 850, 121
O'Malley-James, J. T., Greaves, J. S., Raven, J. A., & Cockell, C. S. 2015, AsBio, 15, 400
Paxton, B., Bildsten, L., Dotter, A., et al. 2011, ApJS, 192, 3
Paxton, B., Cantiello, M., Arras, P., et al. 2013, ApJS, 208, 4
Paxton, B., Marchant, P., Schwab, J., et al. 2015, ApJS, 220, 15
Paxton, B., Schwab, J., Bauer, E. B., et al. 2018, ApJS, 234, 34
Paxton, B., Smolec, R., Schwab, J., et al. 2019, ApJS, 243, 10
Pepe, F., Cristiani, S., Rebolo, R., et al. 2021, A&A, 645, A96
Ramirez, R. M., & Kaltenegger, L. 2014, ApJ, 797, L25
Robertson, P., Anderson, T., Stefansson, G., et al. 2019, JATIS, 5, 015003
Rothman, L. S., Gordon, I. E., Babikov, Y., et al. 2013, JQRST, 130, 4
Rothman, L. S., Gordon, I. E., Barbe, A., et al. 2009, JQRST, 110, 533
Rothman, L. S., Gordon, I. E., Barber, R. J., et al. 2010, JQRST, 111, 2139
Selsis, F., Kasting, J. F., Levrard, B., et al. 2007, A&A, 476, 1373
Sullivan, P. W., Winn, J. N., Berta-Thompson, Z. K., et al. 2015, ApJ, 809, 77
Truitt, A., & Young, P. A. 2017, ApJ, 835, 87
Truitt, A., Young, P. A., Spacek, A., Probst, L., & Dietrich, J. 2015, ApJ, 804, 145
Tuomi, M., Anglada-Escude, G., Gerlach, E., et al. 2013, A&A, 549, A48
Underwood, D. R., Jones, B. W., & Sleep, P. N. 2003, IJAsB, 2, 289
Valle, G., Dell'Omodarme, M., Prada Moroni, P. G., & Degl'Innocenti, S. 2014, A&A, 567, A133
Williams, J. G., & Benson, G. S. 1971, AJ, 76, 167
Yang, J., Boué, G., Fabrycky, D. C., & Abbot, D. S. 2014, ApJ, 787, L2
Young, P. A., Liebst, K., & Pagano, M. 2012, ApJ, 755, L31
Zhang, T. L., Du, J., Ma, Y. J., et al. 2009, GeoRL, 36, L20203

Chapter 7

The Solar System[1]

7.1 In Situ Data

Planetary science has a deep history of observation and robotic exploration within the solar system, from which models of planetary processes have been constructed (de Pater & Lissauer 2015; Horner et al. 2020a; Kane et al. 2021). The robotic exploration of the solar system has been particularly prolific over the past 60 years as we explored beyond the Earth–Moon system, with flybys of Venus and Mars (Fjeldbo et al. 1966; Neugebauer & Snyder 1966) followed, eventually, by planetary landings (Avduevskij et al. 1971; Keldysh 1977; Hess et al. 1977; Toulmin et al. 1977). At this point in history, our spacecraft have visited each of the solar system planets (Smith et al. 1982, 1989), including orbiters to each of the terrestrial planets (Solomon et al. 2001; Nakamura et al. 2011), in addition to Jupiter and Saturn (Belton et al. 1996; Spencer et al. 2006). We have also sent spacecraft to several solar system dwarf planets (Russell & Raymond 2011; Stern et al. 2015) and smaller bodies (Krankowsky et al. 1986; Fujiwara et al. 2006; Glassmeier et al. 2007). The detailed observations and *in situ* measurements of solar system bodies provide the basis of fundamental models that describe the origin and evolution of planetary systems, as well as the nature of atmospheric and geological planetary processes (Goldreich & Soter 1966; Guillot 1999; Lodders 2003; Zahnle et al. 2003).

As described in Section 2.5 and Chapter 8, the age of exoplanet discovery and characterization is upon us, with thousands of known exoplanets and plans to measure atmospheric properties. Exoplanets and planetary science currently face the challenge of an efficient and seamless integration of planetary and exoplanetary science, especially regarding the measurables that are prevalent in these respective fields. Planetary science directly studies the atmosphere, geology, and interiors of planets for which we have spatially and temporally resolved and/or in situ data. However, our knowledge of exoplanet properties is usually not directly obtained

[1] Parts of this chapter have been adapted from Kane et al. (2021).

doi:10.1088/2514-3433/ac2aa1ch7

since terrestrial planets are presently invisible to us. Instead, most exoplanet properties are inferred from their impact on the host star's orbit or brightness. Knowledge of stellar astronomy then becomes the baseline needed to understand exoplanetary characteristics. Thus, although exoplanetary science provides statistical insight into planetary architectures and formation scenarios, planetary science provides detailed planetary models that exoplanetary science relies upon for detailed characterization. For example, the observations of atmospheric escape processes for the solar system terrestrial planets plays an important role in understanding the survival and evolution of exoplanet atmospheres (see Sections 3.2 and 4.4; Strangeway et al. 2005; Tian 2015; Dong et al. 2018a; Gronoff et al. 2020; Lammer et al. 2020). Fundamentally, we must recognize that, although the number of known exoplanets is large and will continue to rapidly grow, we will not have access to *in situ* data for an exoplanet within the foreseeable future. The consequence of this sobering thought is that the proper assessment of exoplanet surface conditions will predominantly be inferred from models based on solar system data. Thus, we must consider the solar system to be our primary resource for advancing our understanding of potentially habitable environments, many parsecs away, that we cannot see.

7.2 The Impact Museums

The inner solar system contains two substantial rocky bodies without atmospheres, and of very similar appearance. Mercury (Figure 7.1; left) is the closest planet to the Sun and is in a 3:2 spin–orbit resonance, meaning that it rotates three times on its spin axis for every two orbits around the Sun. Earth's Moon, Luna, (Figure 7.1; right) is tidally locked to the Earth and receding at the rate of $\sim$3.78 cm yr^{-1} due to the transfer of angular momentum and tidal dissipation (Meyers & Malinverno

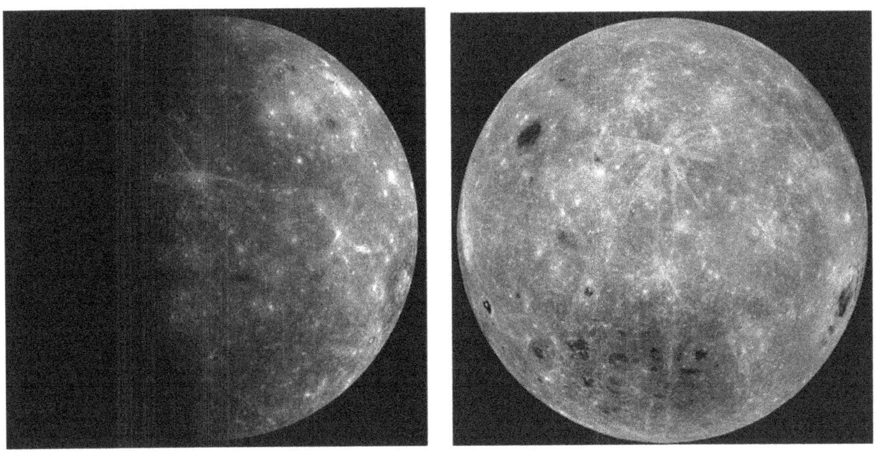

Figure 7.1. Left: image of Mercury taken by the NASA MESSENGER spacecraft. Credit: NASA/Johns Hopkins University Applied Physics Laboratory/Carnegie Institution of Washington. Right: image of the far side of the Moon taken by Lunar Reconnaissance Orbiter spacecraft. Credit: NASA/Goddard Space Flight Center/Arizona State University.

2018). However, although these bodies appear similar, they have quite different origins and interiors.

The orbital reconnaissance of the inner solar system planets by the MESSENGER (MErcury Surface, Space ENvironment, GEochemistry and Ranging) spacecraft revealed that Mercury has experienced sustained, widespread effusive volcanism for the first quarter of its life. Interior cooling and global contraction outpaced melt production and the prevailing stress state shut off major volcanic activity around 3.5 Gyr ago (Byrne et al. 2018). Mercury also has a remarkably thin silicate portion, possibly caused by naturally accreting with much more iron than silica relative to other inner solar system bodies (Solomon et al. 2018). Alternatively, collisions may have stripped off outer layers of the planet while it was forming (Benz et al. 2007; Asphaug & Reufer 2014; Jackson et al. 2018).

A common theme for Mercury and the Moon is the record of past impacts, dating back to much earlier in the solar system history than many other large terrestrial bodies whose impact features have been obscured by erosion and geological activity. MESSENGER observations showed that Mercury has experienced significant impact bombardment, but also is surprisingly volatile rich, especially in sulfur and carbon, for a rocky planet so close to its parent star (Vander Kaaden et al. 2017). The surface of Mercury even contains evidence for ongoing sublimation of a volatile-rich crust (Blewett et al. 2018). However, impact-stripping of Mercury, if it occurred, would have taken place very early in its history since the oldest terrain on the planet is ~4.1 Gyr old (Marchi et al. 2013). Much of the earlier history is likely buried by major volcanic and impact ejecta deposits (Solomon et al. 2018).

The formation of the Moon is thought to have been a stochastic event, the result of a giant collision between "proto-Earth" and a Mars-sized object (colloquially referred to as "Theia"; Benz et al. 1986; Canup & Asphaug 2001; Reufer et al. 2012). However, such stochastic events do not necessarily mean that analog Earth–Moon systems are rare, since a high density of planetesimals in the early stages of planet formation can lead to frequent collisions (Elser et al. 2011). Additionally, as noted in Section 5.4, studies regarding the role of the Moon in stabilizing the spin axis of the Earth (Laskar et al. 1993; Lissauer et al. 2012; Barnes et al. 2016; Ćuk et al. 2016), have suggested that such stabilization may have moderated the Earth's climatic variability (Waltham 2004), and therefore habitability (Williams & Kasting 1997; Spiegel et al. 2009; Heller et al. 2011; Armstrong et al. 2014; Colose et al. 2019).

Like Mercury, the surface of the Moon serves as a time capsule for the impact history of the Earth–Moon system, and how such impacts may have influenced the evolution of Earth's biosphere. Shown in Figure 7.2 is a lunar crater chronology and associated calibration curve, created from studying impact occurrences, locations, velocities, and incidence angles (Le Feuvre & Wieczorek 2011). Calibrations points, based on radiometric ages of impact-melted rocks, are shown as circles. The majority of observed craters on the Moon are estimated to have formed ~3.8–4.1 Gyr ago; a bombardment period that may have truncated early life evolution on Earth (Sleep et al. 1989). Figure 7.2 also suggests a period of sustained impact frequency during 0.5–3.5 Gyr ago, whose source was likely derived from the asteroid population (Neukum et al. 2001). Therefore, the preservation of the impact history on Mercury

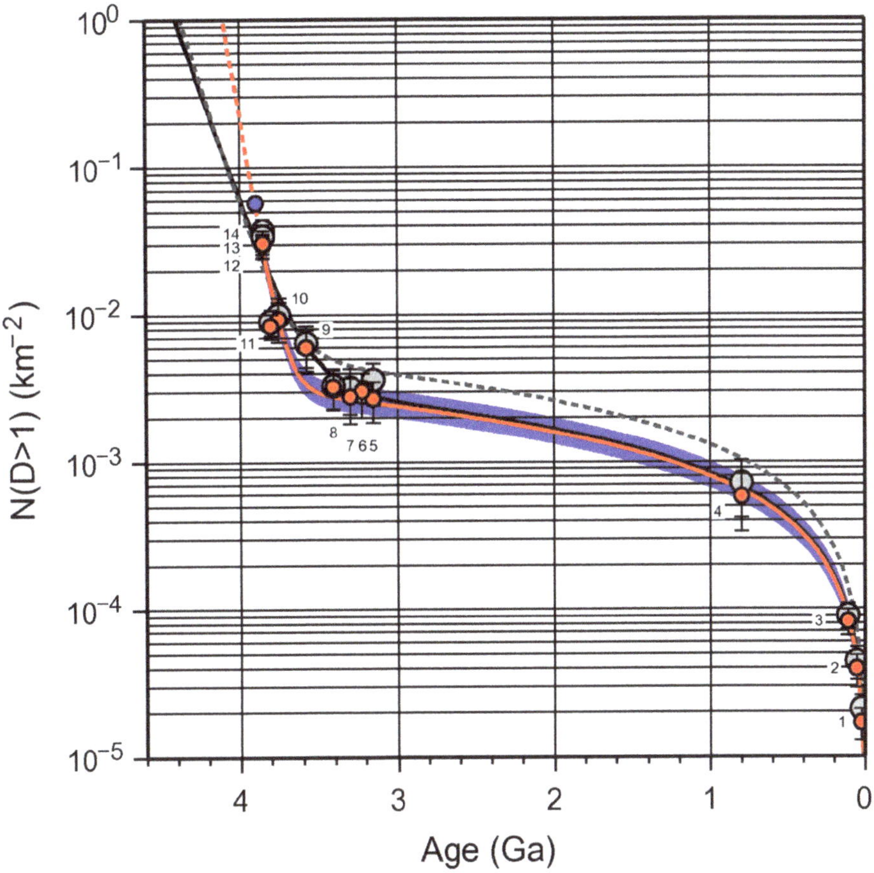

Figure 7.2. Lunar crater chronology calibration curve for crater size as a function of estimated age. The various curves are calibration curves used to extrapolate the impact history of the Moon. Credit: Reprinted from Le Feuvre & Wieczorek (2011), Copyright (2011), with permission from Elsevier.

and the Moon is enormously valuable with respect to planet formation and the evolution of habitability in the inner solar system. Note that there are many other solar system bodies in the outer regions that have limited impact record preservation. However, the low density and icy composition of these bodies tend to result in active geology that can produce surfaces of relatively young ages (see Section 7.6).

7.3 A Tale of Two Siblings

Within the inner solar system, we are fortunate that the largest two terrestrial bodies are similar in size, thus providing two examples of planetary evolution in that size category. Venus and Earth likely formed under similar circumstances, and probably have similar bulk compositions. However, the divergence of their surface and climate evolution is nothing short of extreme, and are generally considered local archetypes of surface hostility and habitability.

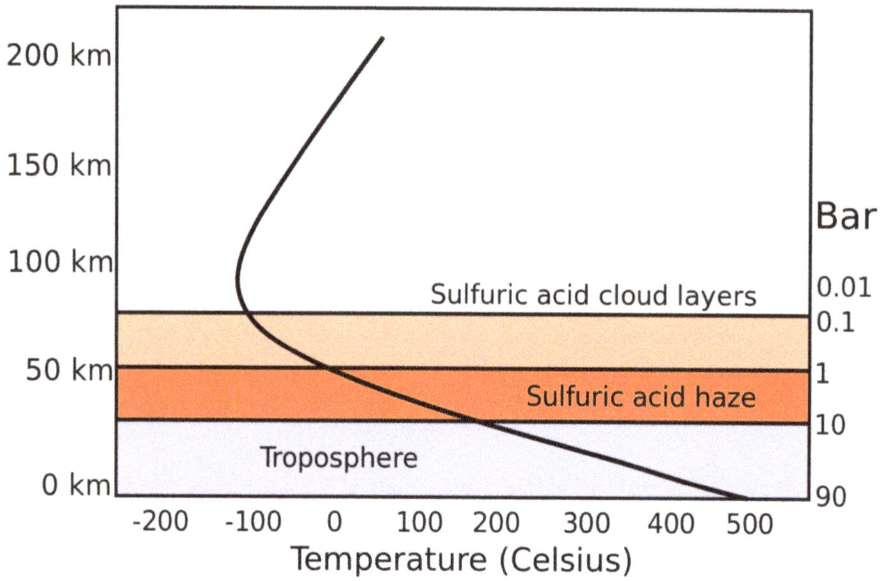

Figure 7.3. Temperature and pressure variation in the Venusian atmosphere as a function of altitude, with the approximate locations of the troposphere and sulfuric acid layers. Credit: Wikimedia: Creative commons license.

Venus is often referred to as Earth's "sibling," because of its similarity to Earth in terms of size and bulk composition. However, it has a ~92-bar atmosphere comprising 96.5% CO_2 and 3.5% N_2, and a surface temperature of ~735 K. Sulfuric acid hazes and clouds persist in the middle atmosphere above the troposphere, whose opaqueness obscures the view of the surface at optical wavelengths. The pressure–temperature profile of the Venusian atmosphere is shown in Figure 7.3, along with approximate altitude locations of sulfuric acid hazes and clouds. Notice that the pressure and temperature at altitudes in the range 50–65 km are similar to those at Earth's surface. The surface of Venus has a surprisingly young average age of ~750 Myr, based on crater counts (Schaber et al. 1992), and may be in a "stagnant lid" state (Herrick 1994), although subduction may still occur today (Davaille et al. 2017; Smrekar et al. 2018; Byrne et al. 2021).

The present atmospheric state of Venus is often explained via a past progression into a runaway greenhouse (Walker 1975) that occurred when the incident solar flux exceeded the limit on outgoing thermal radiation from a moist atmosphere (Komabayasi 1967; Ingersoll 1969; Nakajima et al. 1992; Goldblatt & Watson 2012), evaporating any oceans present (see Sections 3.3 and 6.3). Such a scenario is predicated on the assumption of past oceans on the surface of Venus (Kasting et al. 1984; Grinspoon 1993), although water vapor in the atmosphere may never have condensed on the surface (Hamano et al. 2013). The creation of a moist greenhouse environment can lead to water loss by hydrogen escape which, for Venus, is evident in a high D/H relative to Earth (Donahue et al. 1982; de Bergh et al. 1991). Complete water loss for Earth-equivalent oceans would take a few hundred million years (Watson et al. 1981; Kasting et al. 1984; Zahnle & Kasting 1986; Kasting 1988), depending on extreme ultraviolet (XUV) flux and potential throttled by oxygen accumulation (Wordsworth & Pierrehumbert 2013). Hydration and oxidation of surface rocks (Matsui & Abe 1986) and top-of-the-atmosphere loss processes (Chassefière 1997; Collinson et al. 2016) may have removed any O_2 that was produced by early ocean loss, although it is uncertain how the present atmospheric loss processes would operate for a different (younger) Venus atmosphere, potentially in the presence of a stronger magnetic field (Luhmann et al. 2008; Curry et al. 2015; Persson et al. 2020).

A more recently habitable Venus has been proposed, in which ocean evaporation was not dominated by effects related to secular changes in incident flux (Kane et al. 2020), based on cloud decks forming at the sub-stellar point (Way et al. 2016; Way & Del Genio 2020). Figure 7.4 shows the results from a detailed climate simulation for Venus using solar luminosity values corresponding to 0.715 and 2.9 Gyr ago. The effect of topography and rotation rate are also explored. The top two figures use Venus topography and rotation rate, and demonstrate the cooling effect of cloud formation at the sub-stellar point. The bottom-right figure uses a rotation rate closer to that of Earth, which disrupts the cooling effect at the sub-stellar point and results

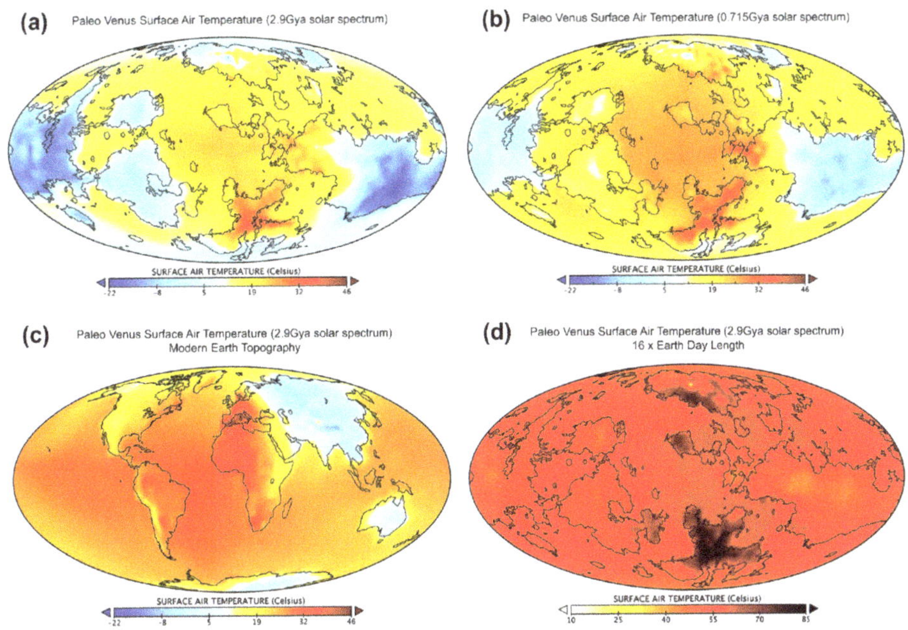

Figure 7.4. The time averaged surface air temperature results from climate simulations for Venus under various incident flux, topography, and rotation rate assumptions. Credit: Way et al. (2016), John Wiley & Sons. © 2016. The Authors. Top left: Paleo-Venus world with the present rotation rate and 2.9 Gyr ago solar luminosity. Top right: same as for (a) but using 0.715 Gyr ago solar luminosity. Bottom left: same as for (a) but using a modern day Earth topography. Bottom right: assuming a 2.9 Gyr ago solar luminosity and a sidereal day 16 times that of modern Earth. Note that the color scale in the bottom-right figure has changed relative to the other three.

in significant warming at the surface. Based on these results, the fundamental difference in climate between Venus and Earth may be more stochastic, and less inevitable, than previously assumed. The explanation for this difference relies upon a deeper understanding of the relative contributions of CO_2 outgassing compared with the physical properties of clouds that determine albedo. Note also that a possible loss of plate tectonics on Venus (relative to Earth) would have played a major role in the moderation of atmospheric CO_2 and the ability to maintain temperature surface conditions (see Section 4.3).

The evolutionary history of Venus provides considerable lessons, and perhaps warnings, regarding both Earth evolution and the analysis of terrestrial exoplanet atmospheres more generally (Schaefer & Fegley 2011; Ehrenreich et al. 2012; Lustig-Yaeger et al. 2019). The bias of exoplanet detection techniques toward shorter orbital periods means that the discovery of potential Venus analogs are expected to become the prime targets for detailed follow-up observations (Kane et al. 2013a, 2014; Ostberg & Kane 2019). For example, Kepler-1649b is a potential Venus analog (Angelo et al. 2017); for which climate simulations predict rapid water-loss scenarios and possible progression through a runaway greenhouse (Kane et al. 2018). However, there remain substantial gaps in our knowledge regarding the dynamics and chemistry of middle and deep atmosphere of Venus, which limit the attempts to perform surface environment modeling of terrestrial exoplanets (Forget 2013; Forget & Leconte 2014). Venus thus serves as a possible local template for the expected challenges in characterizing the evolution and atmospheres of terrestrial exoplanets in or near the HZ (Kane et al. 2019; Lustig-Yaeger et al. 2019).

In parallel to the Venusian evolution pathway, Earth's atmosphere has changed considerably through time. The composition of Earth's earliest atmosphere is poorly known, and it's possible that life may have evolved during the Hadean (>4.0 Gyr ago; Nutman et al. 2016). The evolution of life on Earth has resulted in atmospheric abundances of biosignature gases (e.g., O_2, O_3, CH_4, N_2O) that have varied by orders of magnitude through Earth history (Zahnle et al. 2007, 2020; Schwieterman et al. 2018). These changes in abundances have major implications for the detectability and interpretation of the presence of these gases on exoplanets. The top panel of Figure 7.5 shows a simulated spectrum of Earth during the Archean eon (4.0–2.5 Gyr ago; Arney et al. 2016; Schwieterman et al. 2018), during which Earth's atmosphere is thought to have been relatively anoxic (Lyons et al. 2014), though evidence exists for an earlier oxygen-rich atmosphere (O'Brien 2020). The Archean may have experienced a rise in biologically produced methane (Pavlov et al. 2000), possibly producing haze layers similar to Titan (Trainer et al. 2006; Zerkle et al. 2012; Arney et al. 2016), and whose spectroscopic signature would most easily be observed in its effect on the broadband shape at smaller wavelengths (Gao et al. 2020). The presence of CO_2 and CH_4 (both greenhouse gases) in large quantities may have been sufficient to maintain temperature surface conditions in the presence of a fainter, younger Sun, and whose combination can indicate a biological atmospheric disequilibrium (Krissansen-Totton et al. 2018). The start of the Proterozoic (2.3 Gyr to 541 Myr ago) marked the rise of an oxygenated atmosphere, irreversibly altering the redox state of the atmosphere (Lyons et al. 2014; Planavsky et al. 2014; Reinhard et al. 2017). The rise of oxygen also meant the rise of its photochemical byproduct, the UV-blocking ozone layer, with profound implications for the surface habitability of our planet. The bottom panel of Figure 7.5 shows an example representation of how Earth's atmospheric O_2 content has evolved through time, all the way to the present epoch.

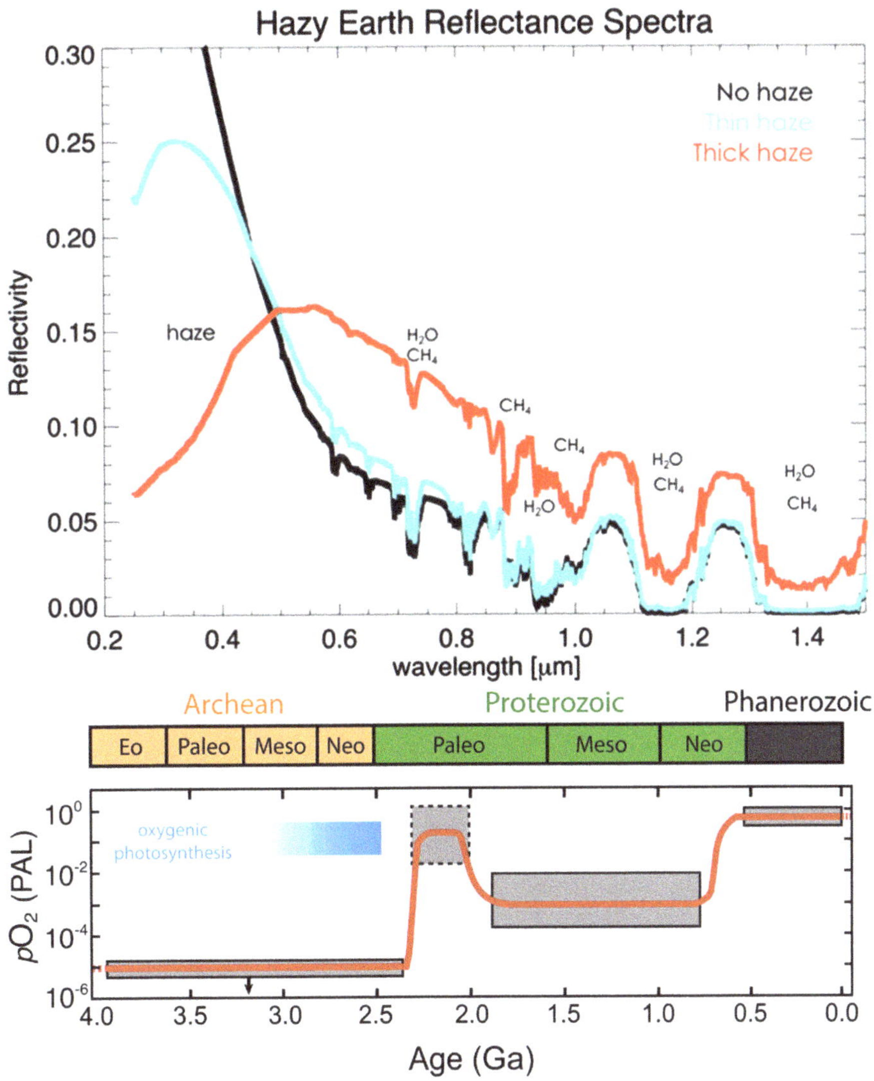

Figure 7.5. Top: spectra of Archean Earth with three different haze thicknesses for atmospheres with 2% CO_2 (Arney et al. 2016; Schwieterman et al. 2018). Bottom: evolution of Earth's atmospheric O_2 content through time, where shaded boxes show approximate ranges based on the latest geochemical proxy records and the curve shows one possible evolutionary trend over time (Lyons et al. 2014; Reinhard et al. 2017; Schwieterman et al. 2018). Credit: Reprinted with permission from Schwieterman et al. (2018), Copyright © Edward W. Schwierterman et al.

The origin of the Earth's water inventory also remains a point of some discussion (Marty 2012; Marty et al. 2016; Dauphas 2017; Peslier et al. 2017; Wu et al. 2018), and various theories for the hydration of Earth are under consideration. Endogenous hydration proposes that the water was accreted from material local to Earth from the solar nebula during formation (Ikoma & Genda 2006), mostly in the form of hydrated silicates (Drake 2005). Early exogenous hydration suggests that volatile material was delivered from beyond the snow line by asteroids and comets during Earth's formation (Morbidelli et al. 2000; Petit et al. 2001). Alternatively, late exogenous hydration suggests that the delivery of water from the outer solar system occurred slightly later, towards the end of Earth's accretion, or even some time after formation (Owen & Bar-Nun 1995). Note that the majority of these models assume that all terrestrial planets received similar amounts of volatile material, including Venus. The consequence of this assumption is that the volatile isotopic abundances of the volatiles should be similar for all of the planets. However, dynamical studies indicate that the terrestrial planets likely received different amounts of material from different reservoirs of volatiles (Owen & Bar-Nun 1996; Raymond et al. 2004; Horner et al. 2009), a result that has been replicated in studies of planet formation around other stars (Ciesla et al. 2015). This finding has implications, for example, for the volatile inventory of Venus and its similarity (or lack thereof) with that of Earth (Way & Del Genio 2020).

7.4 Exploring the Lower Limits

The surface of Mars is extremely diverse, and includes a variety of features, such as impact craters, tectonic structures, and volcanoes and their products (some of which are the largest examples of such features in the solar system; Werner 2009). Mars also preserves evidence for an early Mars (~3.5–4.0 Gyr ago) whose surface contained enough liquid water to carve valley networks (Ehlmann & Edwards 2014; Grotzinger et al. 2014; Galofre et al. 2020). The extent to which early Mars was "warm and wet" relative to its present state is uncertain (Ramirez & Craddock 2018; Galofre et al. 2020). However, in order to sustain surface liquid water, the atmospheric pressure must necessarily have been much higher than the 6 mbar atmosphere present today (Carr 2012). An early dynamo and magnetic field permanently altered the ancient (crust Acuna et al. 1999), and subsequently dramatically reduced before the valley networks formed (Lillis et al. 2008). In essence, Mars represents a case study for a world that was once more geologically active than today (Banerdt et al. 2020; Giardini et al. 2020), and perhaps even once habitable, that underwent a major change in internal and surface properties as its interior cooled and its atmosphere was lost to space (Ehlmann et al. 2016). Indeed, it is possible that Mars represents the lower limit of planetary size where this transition occurs (Ehlmann et al. 2016; Kite 2019), since widespread geological activity ended much earlier on smaller Mercury and the Moon, but continues to the present on Venus and Earth (Byrne 2020).

The tenuous nature of the Martian atmosphere has been considered in the context of exoplanet atmospheric escape (Brain et al. 2016; Dong et al. 2018b). This is particularly important because many exoplanets orbit very closely to their stars and/or their stars exhibit high levels of activity that may be sufficient to strip significant fractions of an atmosphere (see Sections 3.2 and 4.4). Figure 7.6 shows a graphical representation of atmospheric loss from Mars over time. Data from the Mars Atmosphere and Volatile EvolutioN (MAVEN) spacecraft (Jakosky et al. 2015) were used to measure light and heavy isotopes of argon in the Martian atmosphere, and showed that 65% of the original atmospheric argon has been lost to space (Jakosky et al. 2017). In fact, if Mars were closer to the Sun, such as an Earth-like orbit, the atmospheric loss processes would be accelerated, producing an even more hostile surface environment. In the context of exoplanets, the MAVEN data have been used to validate a separate study that suggests 100 bars of atmosphere can be lost from a Mars-like exoplanet orbiting in the HZ of a M-dwarf star over 4 Gyr (Dong et al. 2018b).

In a planetary habitability context, Mars provides insight into a class of planets that could potentially be mistaken for a habitable world. It demonstrates one of the extremes of planetary evolution, starting with habitable conditions, and then rapidly evolving into a borderline or questionable state of habitability (Bishop et al. 2018; Ramirez et al. 2020). The size of Mars also has links to the latter stages of planet formation, with some studies suggesting that its relatively small size is the result of significant collisional attrition, and others arguing that it might be more representative of the oligarchs that precluded the final collisional growth of the two largest terrestrial planets (Izidoro et al. 2015; Brasser et al. 2017; Bromley & Scott 2017). Furthermore, the chaotic changes in the Martian obliquity and orbit combine to give the planet a far greater level of climatic variability than is seen for Venus or Earth (William 1973; Jakosky et al. 1995; Mischna et al. 2013). Finally, studies of the interior are revealing the processes that occur in cases of failed plate tectonics and the planetary outcomes for which such tectonic activity is feasible (Seales & Lenardic 2021).

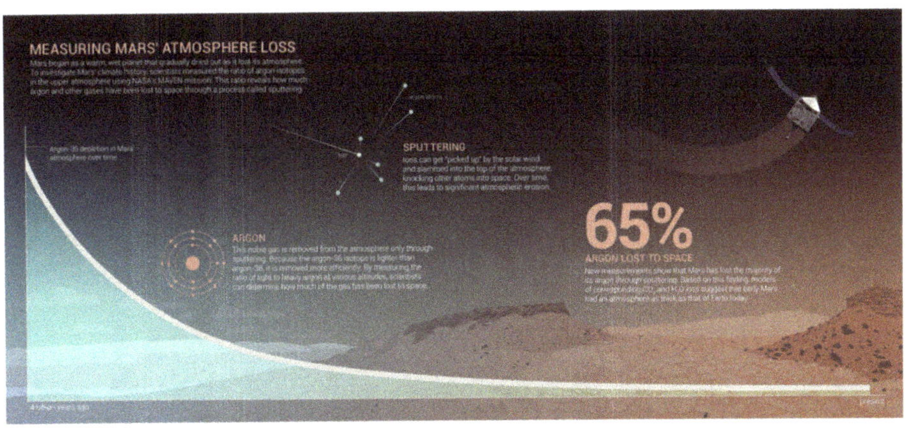

Figure 7.6. A representation of atmospheric loss from Mars over 4 Gyr using data from NASA's MAVEN spacecraft (Jakosky et al. 2017). MAVEN measured light and heavy isotopes of argon in the Martian atmosphere, which serve as tracers for the loss rate as a function of time. Credit: NASA/JPL-Caltech.

7.5 Jupiter: Friend or Foe?

One of the most prominent features of our planetary system is the presence of a giant planet in an orbit ~5.2 AU away from the Sun. Jupiter has a mass that is more than 2.5 times the combined mass of the remaining solar system planets, and thus its gravitational influence can have a profound effect on the stability and evolution of the other solar system bodies. Recent dynamical simulations have shown that the giant planets must have migrated during their formation and evolution, before settling at their current locations. Evidence for their migration include the sculpting of the asteroid belt and orbital distribution of the Jovian Trojans (Morbidelli et al. 2005; Minton & Malhotra 2009), and the excitation of the Plutinos and distribution of objects beyond Neptune (Malhotra 1993, 1995; Hahn & Malhotra 2005; Levison et al. 2008; Lykawka et al. 2009, 2010). However, the extent of that migration, its smoothness or chaoticity, and its timing, remain unclear (Raymond & Alessandro 2002). Migration models include the "Nice Model" (~700 Myr after planet formation; Gomes et al. 2005; Morbidelli et al. 2005), and the relatively early "Grand Tack" migration of Jupiter and Saturn, in which it is suggested that Jupiter might have migrated inward to approach the current orbit of Mars, before tacking outward to reach its current location (Walsh et al. 2011; Raymond et al. 2014; Nesvorný 2018), a process that would have had a significant impact on the hydration of the inner solar system (O'Brien et al. 2014; Raymond & Izidoro 2017). The migration of the giant planets, and Jupiter in particular, therefore likely played a major role in the water inventories of Venus and Earth (Section 7.3) and the size of Mars (Section 7.4).

The presence of giant planets has often been suggested to play a more general role in the habitability of terrestrial planets within the same system (Georgakarakos et al. 2018; Sánchez et al. 2018). Such influence could include offering shielding from an

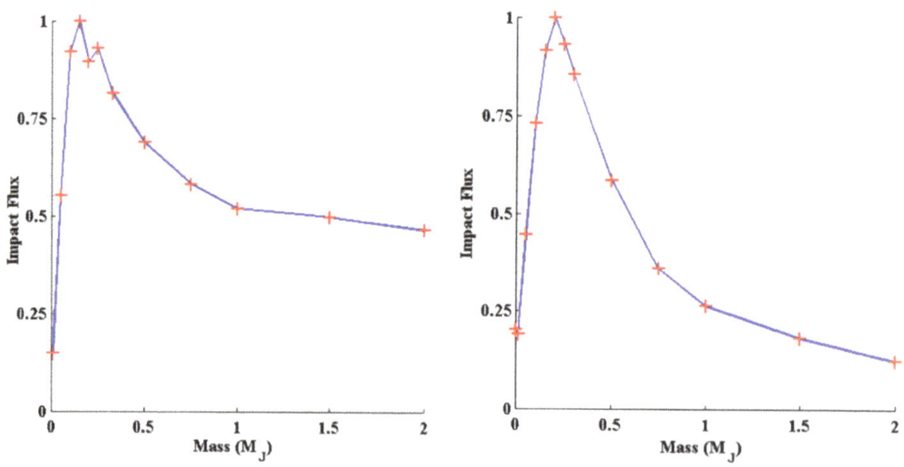

Figure 7.7. The impact rate experienced by the Earth as a result of near-Earth asteroids (left) and Jupiter-family comets (right), as a function of Jupiter's mass (Horner & Jones 2008, 2009; Horner et al. 2020a). The impact fluxes have been normalized to the highest flux experienced by the Earth through the course of the simulations. Credit: Reprinted with permission from Jonti Horner.

impact regime that would otherwise render the planet sterile (Quintana et al. 2016). However, several studies have revealed that the situation is likely far more complex (Horner & Jones 2008, 2009, 2012; Lewis et al. 2013; Grazier 2016; Grazier et al. 2018), with giant planets acting as a source of potential impact threat. Figure 7.7 shows the results of a dynamical simulation that investigated the impact flux of near-Earth asteroids (left) and Jupiter-family comets (right) as a function of Jupiter's mass (Horner & Jones 2008, 2009; Horner et al. 2020a). In both cases, the impact rate due to the influence of Jupiter is higher than if it were not present at all ($M_J = 0$). These calculations include the scattering effects of Jupiter that eject material from the solar system, as well as the perturbing effects that increase Earth impact rates. Thus, the peak impact flux occurs for a planet mass less than 0.25 M_J for which the ejection of material is far exceeded by the perturbations that send objects on an Earth collision course. Although such collisions can cause significant harm to life at the surface, they can also be advantageous, as several models of the origin of Earth's water invoke an exogenous source, requiring the migration of giant planets to deliver volatiles to the inner solar system. Thus, the effects of impacts can have positive (volatile delivery) and negative (extinction events) consequences for terrestrial planets.

As a final note regarding giant planets, and a described in Section 5.3, the perturbing influence of giant planets can have a dramatic effect on system orbital evolution as angular momentum is transferred from the giant planet to terrestrial planets. This, in turn, may play an important role in shaping the climatic variability of potentially habitable worlds by influencing their Milankovitch cycles (Deitrick et al. 2018a, 2018b; Horner et al. 2020b; Kane et al. 2020; Wolf et al. 2020).

7.6 Icy Moons

The solar system contains a large number of moons, 16 of which have a mean radius larger than 500 kms. In particular, icy moons (beyond the snow line) have large water inventories relative to the terrestrial bodies of the inner solar system, due to the availability of volatiles during formation. Section 6.1 describes why sub-surface oceans are not incorporated into the HZ definition. However, these sub-surface ocean environments have the potential to harbor life, particularly when they have additional energy sources not available to a typical planet (see Section 5.4). Given their prevalence within the solar system, substantial effort is being devoted to searching for moons orbiting exoplanets (Hinkel & Kane 2013; Kipping et al. 2013; Heller et al. 2014; Hill et al. 2018; Fox & Wiegert 2021). Furthermore, formation of regular moons, such as those in the Galilean system, may serve as analogs of compact exoplanetary systems in terms of their formation and architectures (Kane et al. 2013b; Makarov et al. 2018; Dobos et al. 2019; Batygin & Morbidelli 2020). Therefore, icy moons and their potentially habitable niches are an important aspect of planetary habitability.

There are numerous questions that remain regarding the wide array of moons in the solar system, including their geology and, in some cases, atmospheres. The icy satellites of the giant planets may serve as small-scale analogs for ocean planets: low-mass, water-rich exoplanets. Ocean planets are a class of terrestrial exoplanets with substantial water layers that may be common throughout the Galaxy (Léger et al.

2004; Raymond et al. 2006; Ehrenreich & Cassan 2007; Sotin et al. 2007), and which are likely to have H_2O inventories at least an order of magnitude greater than Earth's ~0.1% H_2O content. For example, Europa serves as a local example of a sub-surface ocean world, whose water mass fraction is estimated at ~8% (Sohl et al. 2002; Ronnet et al. 2017). Figure 7.8 shows a cross-section of Europa that highlights the processes occurring within the ocean, as well as potentially habitable locations and energy sources. The applicability of Europa to ocean planets depends on many

Figure 7.8. Artist representation of Europa in cross-section showing processes from the seafloor to the surface (not to scale). Inset boxes describe potentially habitable sites, including hydrothermal vents, and regions on and within the ice shell. Also shown are the energy sources at the surface, including charged particles, sunlight, and impacts. Credit: Europa Lander Study 2016 Report. Courtesy of NASA/JPL-Caltech.

factors, such as the climate state, which includes ice-free, partially ice covered, and completely frozen (Tajika 2008; Quick et al. 2020). Ocean planets that are mostly ice-covered may have considerable regions of unfrozen land near their equators or small, equatorial regions of salt-rich water where life could flourish (Del Genio et al. 2019; Paradise et al. 2019; Olson et al. 2020). Additionally, ocean planets in highly eccentric orbits may well also possess substantial amounts of internal energy owing to tidal heating from their host stars (see Section 5.4). However, it is worth noting that, although the premise of habitability described in Section 1.2 assumes the presence of liquid water, it is not safe to assume that increasing the water mass fraction of a planet proportionally increases habitability. A sufficiently large surface water mass can result in high-pressure ice at the mantle boundary, which can truncate geochemical cycles that reduce the flux of bioessential phosphorus (Glaser et al. 2020).

Nonetheless, studying the interiors, tidal properties, and evolution of icy moons may provide similar key insights into the properties of ocean planets (Ehrenreich & Cassan 2007; Sotin et al. 2007; Henning & Hurford 2014; Vance et al. 2007; Noack et al. 2016; Luger et al. 2017; Yang et al. 2017; Barr et al. 2018; Journaux et al. 2020). Owing to their similar internal structures, geophysical processes operating on ocean planets with ice-covered surfaces may be similar to geophysical processes operating on the moons of the giant planets, and may include ice tectonics (Fu et al. 2010; Levi et al. 2014; Hurford et al. 2020), and cryovolcanism (Levi et al. 2013; Quick et al. 2020). Although the specular reflection of starlight, or "glint," on the surfaces of ocean-covered planets will make them fairly easy to detect at visible and near-IR wavelengths (Williams & Gaidos 2018; Robinson et al.2010; Vissier & van de Bult 2015; Lustig-Yaeger et al. 2018) the high albedos of ocean planets with ice-covered surfaces will make them far more detectable than rocky planets in reflected light (Wolf 2017). Many ocean planets may resemble Saturn's largest moon Titan, where the presence of a dense atmosphere allows for the maintenance of liquid at its surface (Lora et al. 2015). With its active methane cycle (Turtle et al. 2011; Dalba et al. 2012; Hörst 2017; Levi & Cohen 2019), Earth-like shorelines (Lunine & Lorenz 2009), diverse geological processes (Jaumann et al. 2009), and the potential for prebiotic chemistry (Neish et al. 2009; He & Smith 2014), Titan serves as an analog for ocean planets that are similar to Earth in nature. Haze in the atmospheres of Titan-like exoplanets could be detected by next-generation space telescopes (Robinson et al. 2014; Checlair et al. 2016; Lora et al.2018), thereby revealing the atmospheric compositions of numerous ocean planets.

References

Acuna, M. H., Connerney, J. E. P., Ness, N. F., et al. 1999, Sci, 284, 790
Angelo, I., Rowe, J. F., Howell, S. B., et al. 2017, AJ, 153, 162
Armstrong, J. C., Barnes, R, Domagal-Goldman, S, et al. 2014, AsBio, 14, 277
Arney, G., Domagal-Goldman, S. D., Meadows, V. S., et al. 2016, AsBio, 16, 873
Asphaug, E, & Reufer, A 2014, NatGe, 7, 564
Avduevskij, V. S., Avduevsky, V. S., Ya, M, et al. 1971, JAtS, 28, 263
Banerdt, W. B., Smrekar, S. E., Banfield, D., et al. 2020, NatGe, 13, 183
Barnes, J. W., Quarles, B., Lissauer, J. J., Chambers, J., & Hedman, M. M. 2016, AsBio, 16, 487

Barr, A. C., Dobos, V., & Kiss, L. L. 2018, A&A, 613, A37
Batygin, K., & Morbidelli, A. 2020, ApJ, 894, 143
Belton, M. J. S., Head, J. W. III, Ingersoll, A. P., et al. 1996, Sci, 274, 377
Benz, W, Anic, A, Horner, J, & Whitby, J. A. 2007, SSRv, 132, 189
Benz, W, Slattery, W. L., & Cameron, A. G. W. 1986, Icar, 66, 515
Bishop, J. L., Fairén, A. G., Michalski, J. R., et al. 2018, NatAs, 2, 206
Blewett, D. T., Ernst, C. M., Murchie, S. L., & Vilas, F. 2018, Mercury's Hollows (Cambridge: Cambridge Univ. Press), 324
Brain, D. A., Bagenal, F, Ma, Y. J., Nilsson, H, & Stenberg Wieser, G 2016, JGRE, 121, 2364
Brasser, R, Mojzsis, S. J., Matsumura, S, & Ida, S 2017, E&PSL, 468, 85
Bromley, B. C., & Kenyon, S. J. 2017, AJ, 153, 216
Byrne, P. K. 2020, NatAs, 4, 321
Byrne, P. K., Ghail, R. C., Celâl Şengör, A. M., et al. 2021, PNAS, 118, 2025919118
Byrne, P. K., Klimczak, C., & Celal Sengor, A. M. 2018, The Tectonic Character of Mercury (Cambridge: Cambridge Univ. Press), 249
Canup, R. M., & Asphaug, E. 2001, Natur, 412, 708
Carr, M. H. 2012, RSPTA, 370, 2193
Chassefière, E 1997, Icar, 126, 229
Checlair, J., McKay, C. P., & Imanaka, H. 2016, P&SS, 129, 1
Ciesla, F. J., Mulders, G. D., Pascucci, I., & Apai, D. 2015, ApJ, 804, 9
Collinson, G. A., Frahm, R. A., Glocer, A., et al. 2016, GeoRL, 43, 5926
Colose, C. M., Del Genio, A. D., & Way, M. J. 2019, ApJ, 884, 138
Ćuk, M., Hamilton, D. P., Lock, S. J., & Stewart, S. T. 2016, Natur, 539, 402
Curry, S. M., Luhmann, J., Ma, Y., et al. 2015, P&SS, 115, 35
Dalba, P. A., Buratti, B. J., Brown, R. H., et al. 2012, ApJ, 761, L24
Dauphas, N 2017, Natur, 541, 521
Davaille, A, Smrekar, S. E., & Tomlinson, S 2017, NatGe, 10, 349
de Bergh, C., Bezard, B., Owen, T., et al. 1991, Sci, 251, 547
de Pater, I., & Lissauer, J. J. 2015, Planetary Sciences (Cambridge: Cambridge Univ. Press)
Deitrick, R., Barnes, R., Bitz, C., et al. 2018a, AJ, 155, 266
Deitrick, R., Barnes, R., Quinn, T. R., et al. 2018b, AJ, 155, 60
Del Genio, A. D., Way, M. J., Amundsen, D. S., et al. 2019, AsBio, 19, 99
Dobos, V., Barr, A. C., & László, L. K. 2019, A&A, 624, A2
Donahue, T. M., Hoffman, J. H., Hodges, R. R., & Watson, A. J. 1982, Sci, 216, 630
Dong, C., Bougher, S. W., Ma, Y., et al. 2018a, JGRA, 123, 6639
Dong, C., Lee, Y., Ma, Y., et al. 2018b, ApJ, 859, L14
Drake, M. J. 2005, M&PS, 40, 519
Ehlmann, B. L., Anderson, F. S., Andrews-Hanna, J., et al. 2016, JGRE, 121, 1927
Ehlmann, B. L., & Edwards, C. S. 2014, AREPS, 42, 291
Ehrenreich, D., & Cassan, A. 2007, AN, 328, 789
Ehrenreich, D., Vidal-Madjar, A., Widemann, T., et al. 2012, A&A, 537, L2
Elser, S, Moore, B, Stadel, J, & Morishima, R 2011, Icar, 214, 357
Fjeldbo, G., Fjeldbo, W. C., & Von Eshleman, R. 1966, Sci, 153, 1518
Forget, F., & Leconte, J. 2014, RSPTA, 372, 20130084
Forget, F. 2013, IJAsB, 12, 177
Fox, C., & Wiegert, P. 2021, MNRAS, 501, 2378

Fu, R., O'Connell, R. J., & Sasselov, D. D. 2010, ApJ, 708, 1326
Fujiwara, A., Kawaguchi, J., Yeomans, D. K., et al. 2006, Sci, 312, 1330
Gao, P., Thorngren, D. P., Lee, G. K. H., et al. 2020, NatAs, 4, 951
Georgakarakos, N., Eggl, S., & Dobbs-Dixon, I. 2018, ApJ, 856, 155
Giardini, D., Lognonné, P., Banerdt, W. B., et al. 2020, NatGe, 13, 205
Glaser, D. M., Hartnett, H. E., Desch, S. J., et al. 2020, ApJ, 893, 163
Glassmeier, K.-H., Boehnhardt, H., Koschny, D., Kührt, E., & Richter, I. 2007, SSRv, 128, 1
Goldblatt, C, & Watson, A. J. 2012, RSPTA, 370, 4197
Goldreich, P., & Soter, S. 1966, Icar, 5, 375
Gomes, R., Levison, H. F., Tsiganis, K., & Morbidelli, A. 2005, Natur, 435, 466
Galofre, A. G., Jellinek, A. M., & Osinski, G. R. 2020, NatGe, 13, 663
Grazier, K. R. 2016, AsBio, 16, 23
Grazier, K. R., Castillo-Rogez, J. C., & Horner, J. 2018, AJ, 156, 232
Grinspoon, D. H. 1993, Natur, 363, 428
Gronoff, G., Arras, P., Baraka, S., et al. 2020, JGRA, 125, e27639
Grotzinger, J. P., Sumner, D. Y., Kah, L. C., et al. 2014, Sci, 343, 1242777
Guillot, T. 1999, Sci, 286, 72
Hahn, J. M., & Malhotra, R. 2005, AJ, 130, 2392
Hamano, K., Abe, Y., & Genda, H. 2013, Natur, 497, 607
He, C., & Smith, M. A. 2014, Icar, 238, 86
Heller, R, Leconte, J, & Barnes, R 2011, A&A, 528, A27
Heller, R., Williams, D., Kipping, D., et al. 2014, AsBio, 14, 798
Henning, W. G., & Hurford, T. 2014, ApJ, 789, 30
Herrick, R. R. 1994, Geo, 22, 703
Hess, S. L., Henry, R. M., Leovy, C. B., Ryan, J. A., & Tillman, J. E. 1977, JGR, 82, 4559
Hill, M. L., Kane, S. R., Seperuelo Duarte, E, et al. 2018, ApJ, 860, 67
Hinkel, N. R., & Kane, S. R. 2013, ApJ, 774, 27
Horner, J., & Jones, B. W. 2008, IJAsB, 7, 251
Horner, J., & Jones, B. W. 2009, IJAsB, 8, 75
Horner, J., & Jones, B. W. 2012, IJAsB, 11, 147
Horner, J., Kane, S. R., Marshall, J. P., et al. 2020a, PASP, 132, 102001
Horner, J., Mousis, O., Petit, J. M., & Jones, B. W. 2009, P&SS, 57, 1338
Horner, J., Vervoort, P., Kane, S. R., et al. 2020b, AJ, 159, 10
Hörst, S. M. 2017, JGRE, 122, 432
Hurford, T. A., Henning, W. G., Maguire, R., et al. 2020, Icar, 338, 113466
Ikoma, M., & Genda, H. 2006, ApJ, 648, 696
Ingersoll, A. P. 1969, JAtS, 26, 1191
Izidoro, A., Raymond, S. N., Morbidelli, A., & Winter, O. C. 2015, MNRAS, 453, 3619
Jackson, A. P., Gabriel, T. S. J., & Asphaug, E. I. 2018, MNRAS, 474, 2924
Jakosky, B. M., Lin, R. P., Grebowsky, J. M., et al. 2015, SSRv, 195, 3
Jakosky, B. M., Slipski, M., Benna, M., et al. 2017, Sci, 355, 1408
Jakosky, B. M., Henderson, B. G., & Mellon, M. T. 1995, JGR, 100, 1579
Jaumann, R., Clark, R. N., Nimmo, F., et al. 1995, JGR, 100, 1579
Journaux, B., Kalousová, K., Sotin, C., et al. 2020, SSRv, 216, 7
Kane, S. R., Barclay, T., & Gelino, D. M. 2013a, ApJ, 770, L20
Kane, S. R., Ceja, A. Y., Way, M. J., & Quintana, E. V. 2018, ApJ, 869, 46

Kane, S. R., Hinkel, N. R., & Raymond, S. N. 2013b, AJ, 146, 122
Kane, S. R., Kopparapu, R. K., & Domagal-Goldman, S. D. 2014, ApJ, 794, L5
Kane, S. R., Arney, G., Crisp, D., et al. 2019, JGRE, 124, 2015
Kane, S. R., Arney, G. N., Byrne, P. K., et al. 2021, JGRE, 126, e06643
Kane, S. R., Vervoort, P., Horner, J., & Pozuelos, F. J. 2020, PSJ, 1, 42
Kasting, J. F. 1988, Icar, 74, 472
Kasting, J. F., Pollack, J. B., & Ackerman, T. P. 1984, Icar, 57, 335
Keldysh, M. V. 1977, Icar, 30, 605
Kipping, D. M., Forgan, D., Hartman, J., et al. 2013, ApJ, 777, 134
Kite, E. S. 2019, SSRv, 215, 10
Komabayasi, M. 1967, JMeSJ, 45, 137
Krankowsky, D., Lammerzahl, P., Herrwerth, I., et al. 1986, Natur, 321, 326
Krissansen-Totton, J., Olson, S., & Catling, D. C. 2018, SciA, 4, eaao5747
Lammer, H., Scherf, M., Kurokawa, H., et al. 2020, SSRv, 216, 74
Laskar, J., Joutel, F., & Robutel, P. 1993, Natur, 361, 615
Le Feuvre, M., & Wieczorek, M. A. 2011, Icar, 214, 1
Léger, A., Selsis, F., Sotin, C., et al. 2004, Icar, 169, 499
Levi, A., & Cohen, R. E. 2019, ApJ, 882, 71
Levi, A., Sasselov, D., & Podolak, M. 2013, ApJ, 769, 29
Levi, A., Sasselov, D., & Podolak, M. 2014, ApJ, 792, 125
Levison, H. F., Morbidelli, A., Van Laerhoven, C., Gomes, R., & Tsiganis, K. 2008, Icar, 196, 258
Lewis, A. R., Quinn, T., & Kaib, N. A. 2013, AJ, 146, 16
Lillis, R. J., Frey, H. V., & Manga, M. 2008, GeoRL, 35, L14203
Lissauer, J. J., Barnes, J. W., & Chambers, J. E. 2012, Icar, 217, 77
Lodders, K. 2003, ApJ, 591, 1220
Lora, J. M., Kataria, T., & Gao, P. 2018, ApJ, 853, 58
Lora, J. M., Lunine, J. I., & Russell, J. L. 2015, Icar, 250, 516
Luger, R., Sestovic, M., Kruse, E., et al. 2017, NatAs, 1, 0129
Luhmann, J. G., Fedorov, A., Barabash, S., et al. 2008, JGRE, 113, E00B04
Lunine, J. I., & Lorenz, R. D. 2009, AREPS, 37, 299
Lustig-Yaeger, J., Meadows, V. S., & Andrew, A. P. 2019, ApJ, 887, L11
Lustig-Yaeger, J., Meadows, V. S., Mendoza, G. T., et al. 2018, AJ, 156, 301
Lykawka, P. S., Horner, J., Jones, B. W., & Mukai, T 2009, MNRAS, 398, 1715
Lykawka, P. S., Horner, J., Jones, B. W., & Mukai, T. 2010, MNRAS, 404, 1272
Lyons, T. W., Reinhard, C. T., & Planavsky, N. J. 2014, Natur, 506, 307
Makarov, V. V., Berghea, C. T., & Efroimsky, M. 2018, ApJ, 857, 142
Malhotra, R. 1993, Natur, 365, 819
Malhotra, R. 1995, AJ, 110, 420
Marchi, S., Chapman, C. R., Fassett, C. I., et al. 2013, Natur, 499, 59
Marty, B. 2012, E&PSL, 313, 56
Marty, B., Avice, G., Sano, Y., et al. 2016, E&PSL, 441, 91
Matsui, T., & Abe, Y. 1986, Natur, 319, 303
Meyers, S. R., & Malinverno, A. 2018, PNAS, 115, 6363
Minton, D. A., & Malhotra, R. 2009, Natur, 457, 1109
Mischna, M. A., Baker, V., Milliken, R., Richardson, M., & Lee, C. 2013, JGRE, 118, 560

Morbidelli, A, Chambers, J, Lunine, J. I., Petit, J. M., Robert, F, Valsecchi, G. B., & Cyr, K. E. 2000, M&PS, 35, 1309
Morbidelli, A., Levison, H. F., Tsiganis, K., & Gomes, R. 2005, Natur, 435, 462
Nakajima, S., Hayashi, Y.-Y., & Abe, Y. 1992, JAtS, 49, 2256
Nakamura, M., Imamura, T., Ishii, N., et al. 2011, EP&S, 63, 443
Neish, C. D., Somogyi, Á., Lunine, J. I., & Smith, M. A. 2009, Icar, 201, 412
Nesvorný, D. 2018, ARA&A, 56, 137
Neugebauer, M., & Snyder, C. W. 1966, JGR, 71, 4469
Neukum, G., Ivanov, B. A., & Hartmann, W. K. 2001, SSRv, 96, 55
Noack, L., Höning, D., Rivoldini, A., et al. 2016, Icar, 277, 215
Nutman, A. P., Bennett, V. C., Friend, C. R. L., van Kranendonk, M. J., & Chivas, A. R. 2016, Natur, 537, 535
O'Brien, D. P., Walsh, K. J., Morbidelli, A., Raymond, S. N., & Mandell, A. M. 2014, Icar, 239, 74
O'Brien, H. 2020, NatGe, 13, 576
Olson, S. L., Jansen, M., & Abbot, D. S. 2020, ApJ, 895, 19
Ostberg, C., & Kane, S. R. 2019, AJ, 158, 195
Owen, T., & Bar-Nun, A. 1996, EM&P, 72, 425
Owen, T., & Bar-Nun, A. 1995, Icar, 116, 215
Paradise, A., Menou, K., Valencia, D., & Lee, C. 2019, JGRE, 124, 2087
Pavlov, A. A., Kasting, J. F., Brown, L. L., Rages, K. A., & Freedman, R. 2000, JGR, 105, 11981
Persson, M., Futaana, Y., Ramstad, R., et al. 2020, JGRE, 125, e06336
Peslier, A. H., Schönbächler, M., Busemann, H., & Karato, S.-I. 2017, SSRv, 212, 743
Petit, J.-M., Morbidelli, A., & Chambers, J. 2001, Icar, 153, 338
Planavsky, N. J., Reinhard, C. T., Wang, X., et al. 2014, Sci, 346, 635
Quick, L. C., Roberge, A., Mlinar, A. B., & Hedman, M. M. 2020, PASP, 132, 084402
Quintana, E. V., Barclay, T., Borucki, W. J., Rowe, J. F., & Chambers, J. E. 2016, ApJ, 821, 126
Ramirez, R. M., & Craddock, R. A. 2018, NatGe, 11, 230
Ramirez, R. M., Craddock, R. A., & Usui, T. 2020, JGRE, 125, e06160
Raymond, S. N., Kokubo, E., Morbidelli, A., Morishima, R., & Walsh, K. J. 2014, in Protostars and Planets VI, Terrestrial Planet Formation at Home and Abroad, ed. H. Beuther, R. S. Klessen, C. P. Dullemond, & T. Henning, 595
Raymond, S. N., & Izidoro, A. 2017, Icar, 297, 134
Raymond, S. N., Mandell, A. M., & Sigurdsson, S. 2006, Sci, 313, 1413
Raymond, S. N., & Alessandro, M. 2002, arXiv:2002.05756
Raymond, S. N., Quinn, T., & Lunine, J. I. 2004, Icar, 168, 1
Reinhard, C. T., Olson, S. L., Schwieterman, E. W., & Lyons, T. W. 2017, AsBio, 17, 287
Reufer, A., Meier, M. M. M., Benz, W., & Wieler, R. 2012, Icar, 221, 296
Robinson, T. D., Maltagliati, L., Marley, M. S., & Fortney, J. J. 2014, PNAS, 111, 9042
Robinson, T. D., Meadows, V. S., & Crisp, D. 2010, ApJ, 721, L67
Ronnet, T., Mousis, O., & Vernazza, P. 2017, ApJ, 845, 92
Russell, C. T., & Raymond, C. A. 2011, SSRv, 163, 3
Sánchez, M. B., de Elía, G. C., & Darriba, L. A. 2018, MNRAS, 481, 1281
Schaber, G. G., Strom, R. G., Moore, H. J., et al. 1992, JGR, 97, 13257
Schaefer, L., & Fegley, B. Jr 2011, ApJ, 729, 6
Schwieterman, E. W., Kiang, N. Y., Parenteau, M. N., et al. 2018, AstBio, 18, 663
Seales, J., & Lenardic, A. 2021, Icar, 367, 114560

Sleep, N. H., Zahnle, K. J., Kasting, J. F., & Morowitz, H. J. 1989, Natur, 342, 139
Smith, B. A., Soderblom, L., Batson, R. M., et al. 1982, Sci, 215, 504
Smith, B. A., Soderblom, L. A., Banfield, D., et al. 1989, Sci, 246, 1422
Smrekar, S. E., Davaille, A., & Sotin, C. 2018, SSRv, 214, 88
Sohl, F., Spohn, T., Breuer, D., & Nagel, K. 2002, Icar, 157, 104
Solomon, S. C., Nittler, L. R., & Anderson, B. J. 2018, Mercury: The View after MESSENGER (Cambridge: Cambridge Univ. Press)
Solomon, S. C., McNutt, R. L., Gold, R. E., et al. 2001, P&SS, 49, 1445
Sotin, C., Grasset, O., & Mocquet, A. 2007, Icar, 191, 337
Spencer, J. R., Pearl, J. C., Segura, M., et al. 2006, Sci, 311, 1401
Spiegel, D. S., Menou, K., & Scharf, C. A. 2009, ApJ, 691, 596
Stern, S. A., Bagenal, F., Ennico, K., et al. 2015, Sci, 350, aad1815
Strangeway, R. J., Ergun, R. E., Su, Y. J., Carlson, C. W., & Elphic, R. C. 2005, JGRA, 110, A03221
Tajika, E. 2008, ApJ, 680, L53
Tian, F. 2015, AREPS, 43, 459
Toulmin, P. III, Baird, A. K., Clark, B. C., et al. 1977, JGR, 82, 4625
Trainer, M. G., Pavlov, A. A., Dewitt, H. L., et al. 2006, PNAS, 103, 18035
Turtle, E. P., Perry, J. E., Hayes, A. G., et al. 2011, Sci, 331, 1414
Vance, S., Harnmeijer, J., Kimura, J., et al. 2007, AsBio, 7, 987
Vander Kaaden, K. E., McCubbin, F. M., Nittler, L. R., et al. 2017, Icar, 285, 155
Visser, P. M., & van de Bult, F. J. 2015, A&A, 579, A21
Walker, J. C. G. 1975, JAtS, 32, 1248
Walsh, K. J., Morbidelli, A., Raymond, S. N., O'Brien, D. P., & Mandell, A. M. 2011, Natur, 475, 206
Waltham, D. 2004, AsBio, 4, 460
William, R. W. 1973, Sci, 181, 260
Watson, A. J., Donahue, T. M., & Walker, J. C. G. 1981, Icar, 48, 150
Way, M. J., & Del Genio, A. D. 2020, JGRE, 125, e06276
Way, M. J., Del Genio, A. D., Kiang, N. Y., et al. 2016, GeoRL, 43, 8376
Werner, S. C. 2009, Icar, 201, 44
Williams, D. M., & Gaidos, E. 2008, Icar, 195, 927
Williams, D. M., & Kasting, J. F. 1997, Icar, 129, 254
Wolf, E. T. 2017, ApJ, 839, L1
Wolf, E. T., Haqq-Misra, J., Kopparapu, R., et al. 2020, JGRE, 125, e06576
Wordsworth, R. D., & Pierrehumbert, R. T. 2013, ApJ, 778, 154
Wu, J., Desch, S. J., Schaefer, L., et al. 2018, JGRE, 123, 2691
Yang, J., Ding, F., Ramirez, R. M., et al. 2017, NatGe, 10, 556
Zahnle, K., Arndt, N., Cockell, C., et al. 2007, SSRv, 129, 35
Zahnle, K., Schenk, P., Levison, H., & Dones, L. 2003, Icar, 163, 263
Zahnle, K. J., & Kasting, J. F. 1986, Icar, 68, 462
Zahnle, K. J., Lupu, R., Catling, D. C., & Wogan, N. 2020, PSJ, 1, 11
Zerkle, A. L., Claire, M. W., Domagal-Goldman, S. D., Farquhar, J., & Poulton, S. W. 2012, NatGe, 5, 359

AAS | IOP Astronomy

Planetary Habitability

Stephen R Kane

Chapter 8

The Exoplanet Opportunity

8.1 The Statistical Hammer

With respect to exoplanets, we live in an age of magnificent discovery. These discoveries are still relatively recent, with the vast majority occurring over the past few decades. As such, we are still grappling with many of the new findings, some of which challenge planetary models that are derived purely from solar system observations. While the solar system remains our best studied example of a planetary system, observations of exoplanets have expanded our horizon to reveal planets and planetary system architectures that are substantially different from our home system. Some of these new planetary types include hot Jupiters, sub-Neptunes, and volatile-rich terrestrial planets. These include systems, some orbiting G dwarfs like our Sun, where multiple planets are found within the equivalent of the orbit of Mercury (Lissauer et al. 2011), speaking to the importance of planetary migration as a fundamental system process (Walsh & Morbidelli 2011; Gillon et al. 2017; Luger et al. 2017c; Ramos et al. 2017; Unterborn et al. 2018b), and also highlighting the likely importance of gravitational interactions and tidal heating. Extrapolations of the probable demographics in the not yet fully explored regions of exoplanet systems also hint at the relative rarity of Jupiters analogs (Zechmeister et al. 2013; Wittenmyer et al. 2016, 2020; Fulton et al. 2021). These planets have helped us better understand fundamental processes such as atmospheric loss and planetary migration, and have served to place our particular planetary architecture in a much broader context of planet formation. As described in Chapter 7, planetary science forms the foundation of planetary system processes, both in terms of data and models, from which exoplanet observables may be interpreted. In turn, exoplanet observables provide vast numbers of exoplanets from which demographic studies can inform the studies of planetary system formation and evolution in general (Clanton & Gaudi 2016; Barclay et al. 2017; Nielsen et al. 2019). Therefore, although there certainly have been and will continue to be individual exoplanet characterization case studies, the primary power of exoplanets is the "statistical

doi:10.1088/2514-3433/ac2aa1ch8

hammer" that is brought to bear on understanding the distribution of planetary parameters, and their influence on terrestrial surface conditions (see Section 2.1).

Among the various planetary properties that have been widely measured, numerous patterns in the distributions of those parameters have begun to take shape. As described in Section 2.5, a prominent focus of the exoplanet community is exploring the mass–radius relationship, especially for terrestrial planets where the range in bulk densities for a given radius can be relatively large compared with giant planets. This is important for interior models, since the bulk density is a key input into modeling compositions and phase transitions (Unterborn et al. 2016). Careful analysis of planetary radii has also revealed a gap in the radius distribution, shown in Figure 8.1. This radius gap, referred to as the "Fulton gap" or "photo-evaporation valley," has been found to be true for orbital periods of less than 100 days, and appears to lie between the super-Earth and sub-Neptune populations. The cause of the radius gap has been explained through the photo-evaporation hypothesis (Lopez & Fortney 2013; Owen & Wu 2013; Fulton et al. 2017), which suggests that planets in the super-Earth category may consist of a "bare" core that has suffered atmospheric stripping within 100 Myr of formation, due to their proximity to the host star. Planets in the sub-Neptune category, on the other hand, received less high-energy radiation and maintained a thick atmospheric envelope. The discovery of this feature in the radius distribution is important for terrestrial planetary habitability because it sheds considerable light on the topic of atmospheric

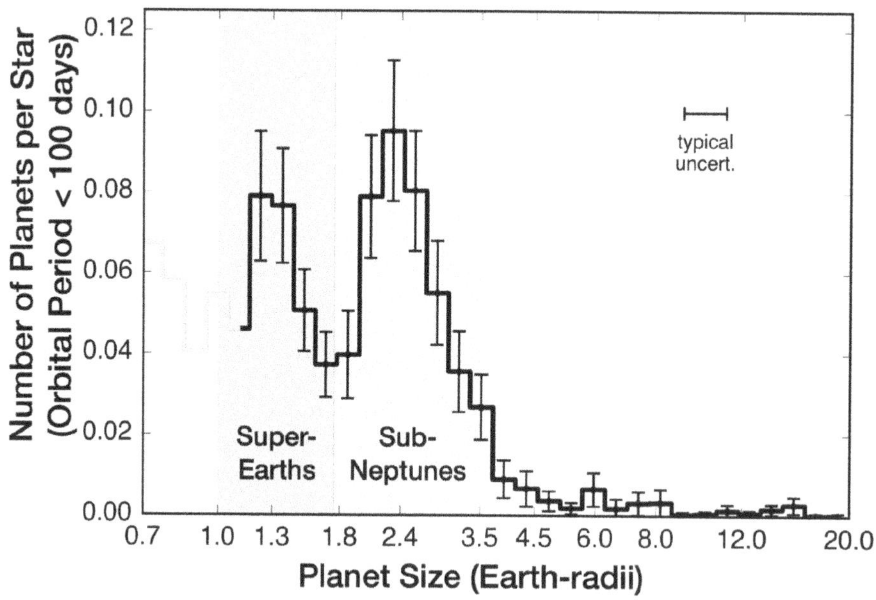

Figure 8.1. Histogram of planetary radii for exoplanets with orbital periods shorter than 100 days. The light gray region of the histogram (radii <1.14 $R_{\oplus}$) suffers from low completeness. The median radius uncertainty is shown in the upper right portion of the plot. Broad definitions of super-Earths and sub-Neptunes are encompassed by the light red and light cyan shaded regions, respectively. Credit: Reproduced from Fulton et al. (2017).

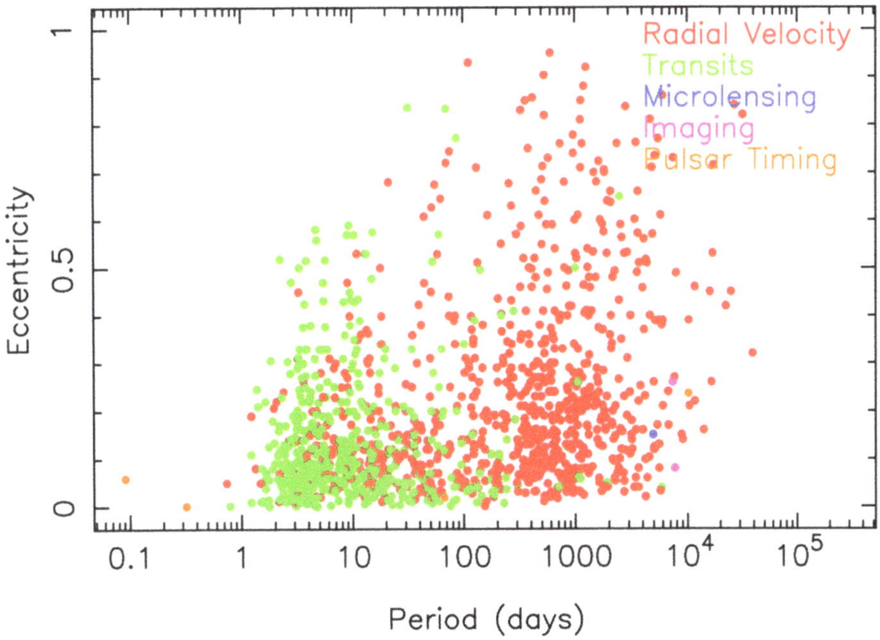

Figure 8.2. Eccentricity and orbital period for the known exoplanets, as of 2021 June 28, shown color-coded by the method of discovery for the individual planets. The data were extracted from the NASA Exoplanet Archive (Akeson et al. 2013).

erosion (see Section 4.4), and the processes that define the limitations of temperate surface sustainability.

A further example of statistical insights to the properties of exoplanets is demonstrated in Figure 8.2; which shows the eccentricity and orbital periods of the known exoplanets, color-coded by the detection technique. Similar to Figure 1.3, the data are current as of 2021 June 28 and were extracted from the NASA Exoplanet Archive[1] (Akeson et al. 2013). Notice that Figures 1.3 and 8.2 reveal the same extreme bias of the transit method toward short orbital periods, discussed further in Section 8.2. Eccentricities are primarily determined through radial velocity measurements which are better suited to characterizing the shape of the orbit (see Section 5.1). As shown in Figure 8.2, eccentricities tend to be low at short orbital periods, but the range of eccentricities rapidly increases at larger orbital periods. This trend in eccentricities has been observed and statistically investigated on numerous occasions (Shen & Turner 2008; Moorhead et al. 2011), and is thought to be primarily caused by tidal dissipation and circularization, in which the tidal forces applied to the planet by the host star act to reduce the orbital energy, thus reducing the eccentricity (Goldreich & Soter 1966; Laskar et al. 2012). By comparing the exoplanet eccentricity distribution with those of binary stars, it has also been found that eccentricities may be indicative of formation scenarios and subsequent scattering events (Jurić &

[1] https://exoplanetarchive.ipac.caltech.edu/

Tremaine 2008; Bowler et al. 2020). With respect to planetary habitability, Sections 5.1 and 5.2 quantify the effect of eccentricity on the energy balance of the planet. However, the finding that eccentricities tend to be lower for terrestrial planets compared with giant planets (Kane et al. 2012; Van Eylen & Albrecht 2015), means that terrestrial planets likely occupy the lower regions of Figure 8.2.

One of the most important applications for the vast number of exoplanet discoveries is the statistical inference of architectural occurrence rates. As mentioned above, estimates of Jupiter analog occurrence rates have shown them to be relatively rare. However, the greatest interest for these analyses has been in determining the rate at which planetary systems harbor a terrestrial planet within the habitable zone (HZ). Such calculations were the primary goal of the Kepler mission (Borucki 2016), whose uniform survey design enabled an unprecedented statistical sample from which these studies could be undertaken. Indeed, many groups have published their calculations (Catanzarite 2011; Dressing & Charbonneau 2013; Gaidos 2013; Kopparapu 2013; Foreman-Mackey et al. 2014; Dressing & Charbonneau 2015; Dulz et al. 2020), producing a quantity referred to as $\eta_{\oplus}$, or "eta Earth," which is the average number of terrestrial planets in the optimistic HZ per star. Figure 8.3 shows

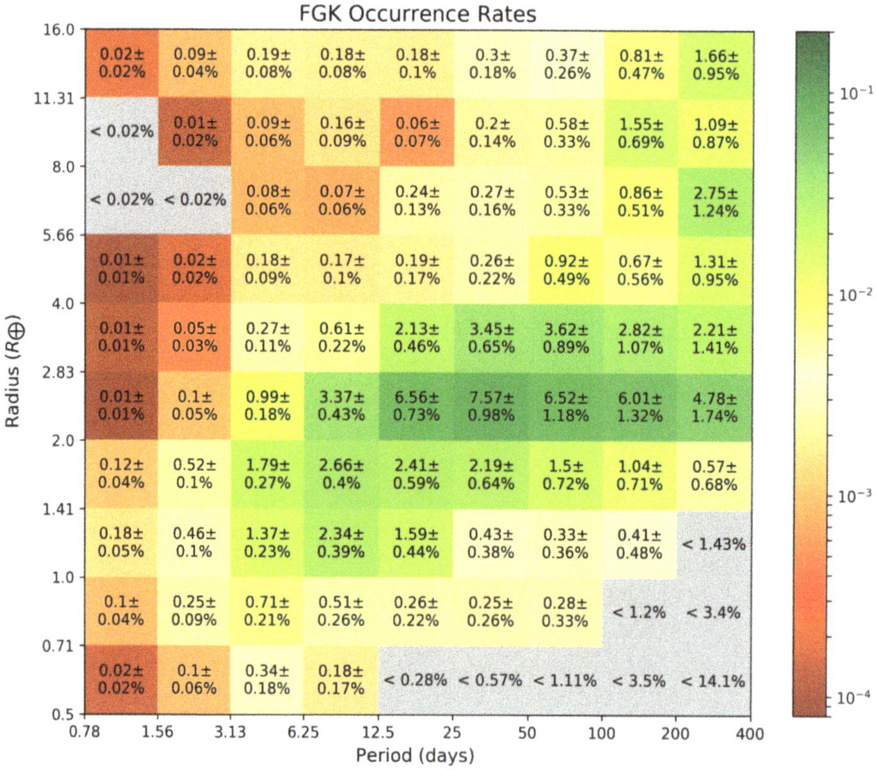

Figure 8.3. Occurrence rate estimates for FGK-type stars using Kepler data, where the number of planets per star are shown as percentages. Bins with no detected planets are in gray, with only the upper limit shown. Credit: Reproduced from Kunimoto & Matthews (2020).

an example of occurrence rate calculations for all of the F, G, and K spectral type stars within the Kepler sample (Kunimoto & Matthews 2020). Typically, the data are divided into bins of planetary radii and orbital period, which are two of the main measurables provided by the transit method. The number of planet discoveries within each bin are corrected for the observational biases of the transit method (see Section 8.2) and the total number of stars observed. Notice that the gray regions of Figure 8.3 indicate parameter space areas of relative incompleteness and so only upper limits are able to be determined. Unsurprisingly, these areas tend to lie in the bottom-right region of the figure due to the difficultly in detecting the signatures of long-period terrestrial planets. The final data release from the Kepler mission provided a robust estimate for $\eta_{\oplus}$ of ~0.58–0.88 for solar-type stars (Bryson et al. 2021). It is expected that this estimate will be further refined as exoplanet survey data continue to provide increasing sensitivity to Earth analogs.

8.2 Data Limitations

Given all that exoplanet discoveries have achieved and the characterization that has occurred, it is worth revisiting the factors that influence planetary habitability outlined in Chapter 2, and the capabilities of present exoplanet techniques, summarized in Table 2.1. It should be understood that the large exoplanet inventory is not necessarily indicative of how proficient our civilization has become at finding planets, but rather it is a reflection of how common planets are. As occurrence rate calculations have shown, our current detection sensitivity drops dramatically approximately in the area of parameter space where true Earth analogs are expected to reside. Furthermore, we should not forget that the two most prolific detection techniques, radial velocities and transits, are indirect methods, and so are used to infer, rather than observe, the presence of the planet. Consequently, the information that we glean from discovered planets is largely limited to orbital parameters and rudimentary bulk properties, such as size and radius. As outlined in Section 2.1 and the "planets are hard" diagram shown in Figure 2.1, planetary habitability is exceptionally complex and many of the required planetary parameters are presently beyond our reach to measure. Therefore, although we strive to assess the habitability of planets we cannot see that are many parsecs distant, the degeneracy of our planetary surface models is an inevitable outcome of our current data limitations. If you consider that vast lack of knowledge that exists regarding the nearest Earth-size planet, Venus (see Section 7.3)—the third brightest object in the sky—then you will understand the scope of the problem.

The observing techniques used to discover and characterize exoplanets are currently most sensitive to planets that are close to their stars (via transits and radial velocities), or very far from their stars (via direct imaging and astrometry). The transit method is exceptionally popular and has utilized several space-based telescopes that are either wholly or partially dedicated to transit exoplanet science. There are good reasons for this popularity, such as the ability to determine the planetary radius, and the potential for carrying out transmission spectroscopy observations that can measure atmospheric composition (see Section 6.4).

However, the transit method has significant disadvantages that should be noted. The transit method, by definition, only measures the planet when it is located at or near inferior conjunction. This means that information regarding the planet outside of the stellar disk crossing event is absent and the shape of the orbit (eccentricity) is relatively difficult to constrain (Kipping 2014; Agol & Deck 2016).

By far, the major limitation of the transit method is that regarding the transit probability, P_t, expressed as follows:

$$P_t = \frac{R_p + R_\star}{a} \approx \frac{R_\star}{a} \tag{8.1}$$

where it is often assumed that $R_p \ll R_\star$. The a^{-1} dependence means that the transit probability undergoes a rapid decline with separation from the host star (Burke 2008; Kane & von Braun 2008). This is demonstrated in Figure 8.4, which plots the transit probability as a function of semimajor axis for the Sun (blue) and TRAPPIST-1 (red). To place this in perspective, the semimajor axes of the solar system planets are shown as vertical dotted lines, and the location of the conservative (light green) and optimistic (dark green) HZ regions are also indicated for both stars. It can be seen that the probability that a solar system planet will

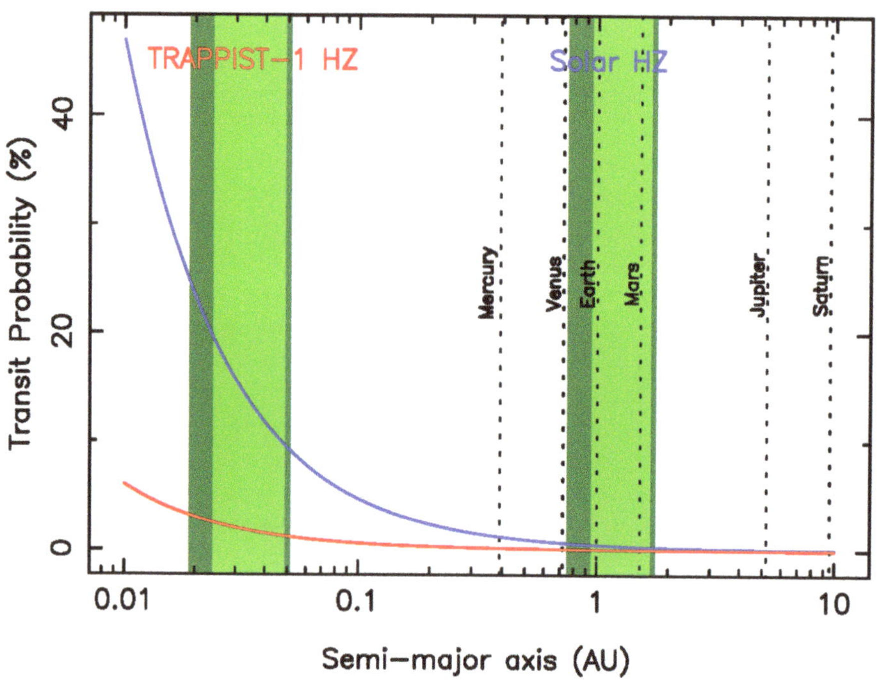

Figure 8.4. Transit probability versus semimajor axis, assuming $R_p \ll R_\star$. The blue line corresponds to the case of a solar radius, and the semimajor axes of solar system planets are shown as vertical dotted lines. The red line corresponds to the case of TRAPPIST-1, whose radius is $R_\star = 0.12\ R_\odot$. The location of the conservative (light green) and optimistic (dark green) HZ regions are shown for both TRAPPIST-1 and the Sun.

transit the Sun from the perspective of a randomly located external observer is extremely low, and is ~0.05% for Earth. Even for the TRAPPIST-1 system, the transit probability of HZ planets is <5%. Integrated over transit probability, the transit method is (currently) simultaneously the most successful and least efficient detection method.

So why does the transit method work at all? Transit surveys overcome the probability issue through brute force, by monitoring many thousands of stars simultaneously to increase the odds of a successful detection. However, the probabilities shown in Figure 8.4 demonstrate that, although transit detections are made, the method fails to detect the vast majority of planets that exist. Unlike other techniques, where improving precision will allow the detection of more planets, the transit method rapidly hits a ceiling whereby non-transiting planets remain invisible, regardless of photometric precision. This is particularly important for nearby systems, whose proximity will enable more detailed direct imaging prospects, and indeed the greatest public excitement. The low transit probability of the TRAPPIST-1 system, which is located ~12 parsec away, means that there is a much closer analog to this system that doesn't transit. Moreover, Figure 8.4 indicates that, even for M dwarf systems, more than 95% of potentially habitable planets do not have orbital inclinations that would result in visible transits. Thus, we can state with very high confidence that the nearest habitable exoplanet does not transit its host star. The transit method is optimally suited for exoplanet statistical studies, such as the occurrence analyses described in Section 8.1, rather than individual characterization.

The radial velocity method measures the gravitational influence of the planet on the host star, and thus is dependent upon the planet mass. Figure 8.5 is similar to Figure 1.3, except that semimajor axis is shown instead of orbital period, and the planet masses are represented in Earth mass units. The dashed lines indicate different levels of measurement precision, ranging from 10 cm s^{-1} to 3 m s^{-1}, and are representative of present radial velocity surveys (Fischer & Anglada-Escude 2016). The measurement precision calculations assume a solar mass host star. Planetary objects that lie above the lines could, in principle, be detected by a survey with the stated precision, depending on the survey duration and the activity of the host star (Saar & Donahue 1997; Wright 2005; Desort et al. 2007). Recall from Section 6.4 that the amplitude of the radial velocity variations that Earth induces on the Sun is ~9 cm s^{-1}, which matches what is shown for the 10 cm s^{-1} in Figure 8.5 since the line passes approximately through 1 Earth mass at 1 AU. Given that radial velocity instrument development is rapidly improving, expectations are high that Earth analogs may indeed be within reach using this technique (Robertson et al. 2019; Petersburg et al. 2020; Pepe et al. 2021).

The radial velocity method does not require an orbit that is aligned with the line of sight and so the entire orbit can be equally sampled, providing detailed information regarding the eccentricity and other Keplerian orbital parameters (see Section 5.1). However, since it is a radial effect that is measured, the amplitude of the radial velocity signal depends on the inclination of the orbit relative to the plane of the sky, which is often unknown. The consequence of this is that the measured mass of the planet is in fact a minimum mass, where the true mass is a factor of $1/(\sin i)$

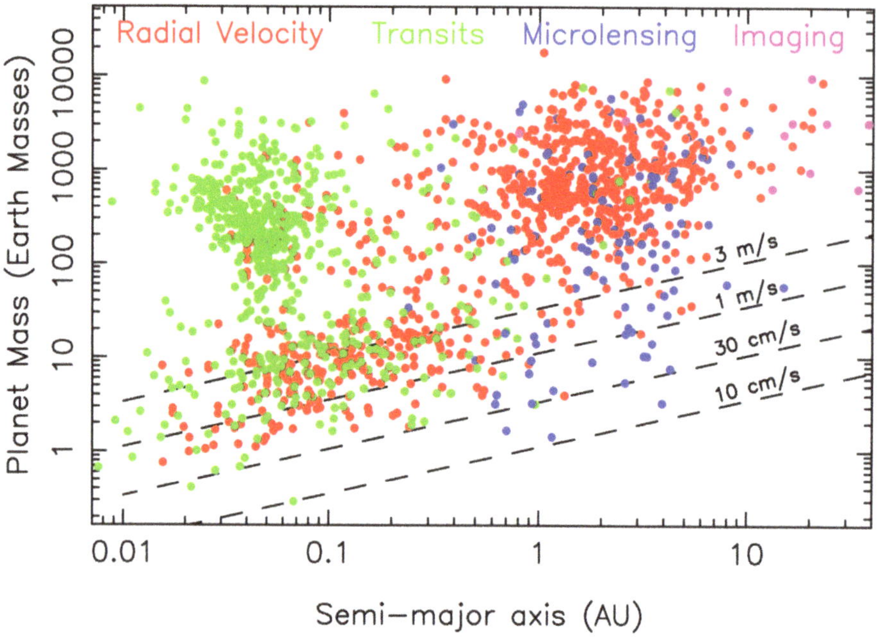

Figure 8.5. Planet mass versus semimajor axis for the known exoplanets, where color-coding is used to identify the method of detection. The data are current as of 2021 June 28, and were extracted from the NASA Exoplanet Archive (Akeson et al. 2013). The dashed lines represent approximate detection limits of varying precision, ranging from 10 cm s^{-1} to 3 m s^{-1}.

higher. This means that for orbital inclinations that are close to face-on ($i = 0°$), the orbiting companion may be stellar in nature rather than planetary. The issue of the unknown inclination can either be solved by astrometric observations or in the highly unlikely event that the planet happens to transit the host star.

Returning to Table 2.1, let us suppose the case of a terrestrial planet that has both a presently measurable transit and radial velocity signal, already a relatively rare occurrence within the broader exoplanet population. The parameters that may be reliably extracted for this case are the planet mass and radius, from which a bulk density may be inferred. We will also measure the orbital period and calculate the semimajor axis, from which the incident flux at the top of the atmosphere may be inferred. Assuming sufficient radial velocity precision, we may also measure the orbital eccentricity, and thus estimate variations in incident flux, if any. If we are fortunate, the host star is either intrinsically bright or close (or both), creating the potential for transmission spectroscopy observations to occur. However, many critical planetary features remain unknown, including atmospheric dynamics, rotation rate, obliquity, or any visual confirmation of continents or surface water. Pessimistically, one may argue that claims regarding habitability with such limited data would be akin to assessing if an unseen person in a different building is alive or dead, having only been told the wattage of their lighting and the radius of their head. However, though we have in many ways only begun to scratch the surface of measuring habitability factors, developing a pathway toward the robust determination of pressure–temperature profiles provides cause for future optimism.

8.3 Notable Exoplanet Examples

With thousands of known exoplanets at the present time, there are numerous cases of planetary systems that harbor at least one planet within the HZ. As of 2021 August 14, the Habitable Zone Gallery[2] reports that there are 143 known planets whose orbits are entirely within the optimistic HZ. These planets are not all terrestrial, and indeed the number of terrestrial HZ planets that may be considered useful for detailed characterization remains relatively small. An outline of five notable exoplanetary systems is provided here. One thing to notice up-front is that all five systems described here are based around M dwarf host stars, whose relatively cool photospheres and low luminosities means that only short period planets will lie in their HZs. This may have several consequences, both positive and negative for habitability, including main sequence lifetime (Section 3.1), stellar activity (Section 3.2), and atmospheric erosion (Section 4.4). However, if continued studies of these cases demonstrate that sustainable habitability is feasible in these environments, then the habitability well may be deep indeed, considering that M dwarfs are the most common stellar type in the known universe.

8.3.1 Kepler-186

As described in Section 8.1, the primary goal of the Kepler mission was to determine the occurrence rate of Earth-size planets within the HZ of their host stars. One of the first Kepler discoveries that contributed to the $\eta_\oplus$ exoplanet inventory was Kepler-186f. The inner four planets, b–e, were announced in a substantial catalog of Kepler planetary candidates (Rowe et al. 2014), whose validation largely stemmed from their detection as multi-planet systems (Lissauer et al. 2012). Further analysis of the Kepler data for Kepler-186 revealed that there was an additional transiting planet, designated planet f (Quintana et al. 2014), with an orbital period of $\sim$129 days, placing it near the outer edge of the star's HZ. The planetary radius was estimated to be relatively small, only 17% larger than the Earth (Quintana et al. 2014; Torres et al. 2015). The size of the planet and location within the HZ are depicted in Figure 8.6, which is an artist's conception of the system with comparison to the solar system. The known extent of the system out to planet f is approximately the same size as the orbit of Mercury. The location of the planetary orbit near the outer edge of the HZ possibly produces a colder climate, resulting in the prospect of large polar ice caps shown in Figure 8.6. Climate models of Kepler-186f have shown that surface temperatures above the freezing point of water can be maintained with 0.5–5 bars of CO_2, depending on the amount of N_2 (Bolmont et al. 2014). The system is an excellent example of a prospective test of planetary habitability for a terrestrial planet in a system architecture that is substantially different from our own. However, the distance to the system is 177.51 ± 0.79 parsec (Kane 2018) and

[2] http://hzgallery.org/

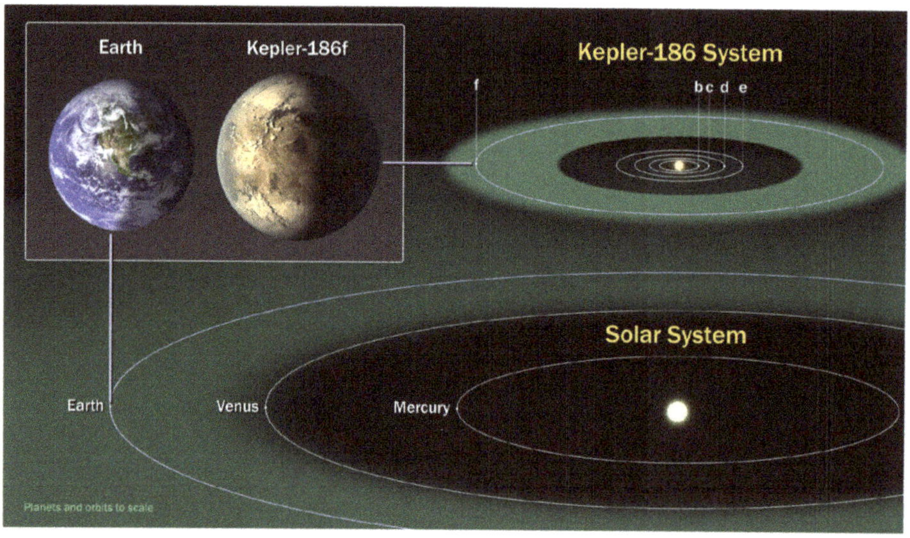

Figure 8.6. Artist's depiction of the Kepler-186 system, showing the size of Kepler-186f relative to Earth, and the extent of the HZ in comparison to the inner solar system (Quintana et al. 2014). Credit: Courtesy of NASA/Ames/SETI Institute/JPL-Caltech.

the host star has a magnitude of $V = 15.138$, unfortunately making it too faint for near-term transmission spectroscopy follow-up observations.

8.3.2 Proxima Centauri

Significant effort has been expended on searching the nearest star system, Alpha Centauri, for exoplanet signatures (Hinkel & Kane 2013; Krishnamurthy et al. 2021). The primary components, A & B, are both similar to the Sun and are in an eccentric orbit ($e = 0.52$) around each other with an orbital period of ∼79.9 years (Hartkopf et al. 2001). Alpha Centauri A & B are located at a distance of 1.34 parsec, or 4.37 light years. Although claims have been made regarding exoplanets around these stars (Dumusque et al. 2012; Hatzes 2013; Demory et al. 2015), no planets have been confirmed at the present time. The third stellar component to the Alpha Centauri system is Proxima Centauri, or Alpha Centauri C, and is the closest star to the Sun at a distance of 1.30 parsec (or 4.24 light years). Unlike the A & B components, Proxima is too faint to be visible to the naked eye, and thus was not discovered until 1915, well after the invention of the telescope. Proxima is gravitationally bound to Alpha Centauri AB, which it orbits with a semimajor axis of ∼13,000 AU (Kamper & Wesselink 1978; Matthews & Gilmore 1993; Kervella et al. 2017). As previously shown in Figure 3.2, Proxima is known to exhibit significant flare activity (Davenport et al. 2016; MacGregor et al. 2018) and has been studied in the context of spectroscopic stellar activity (Robertson et al. 2016).

Precision radial velocity observations of Proxima revealed the presence of a planet in a possibly eccentric 11.2 day orbit, Proxima Centauri b, with an estimated minimum mass of 1.3 $M_\oplus$ (Anglada-Escudé et al. 2016). This ground-breaking discovery means that Proxima Centauri b is, by definition, the closest exoplanet and motivates more detailed studies of our nearest stellar neighbors. A preliminary search for transits of the planet resulted in a claimed detection (Liu et al. 2018), but other observations were able to rule out the possibility that the planet transits (Kipping et al. 2017; Jenkins et al. 2019). Even though the planet doesn't transit, the relatively low measured minimum mass results in a very high probability that the planet is indeed terrestrial (Kane et al. 2017). There has been a tentative detection of a second planet with a semimajor axis of ~1.5 AU (Damasso et al. 2020; Damasso & Del Sordo 2020); but this detection has not been confirmed at the present time.

Although the orbital period of the planet is small, Proxima is also relatively small with low luminosity, and the planetary orbit lies within the HZ. Figure 8.7 shows a

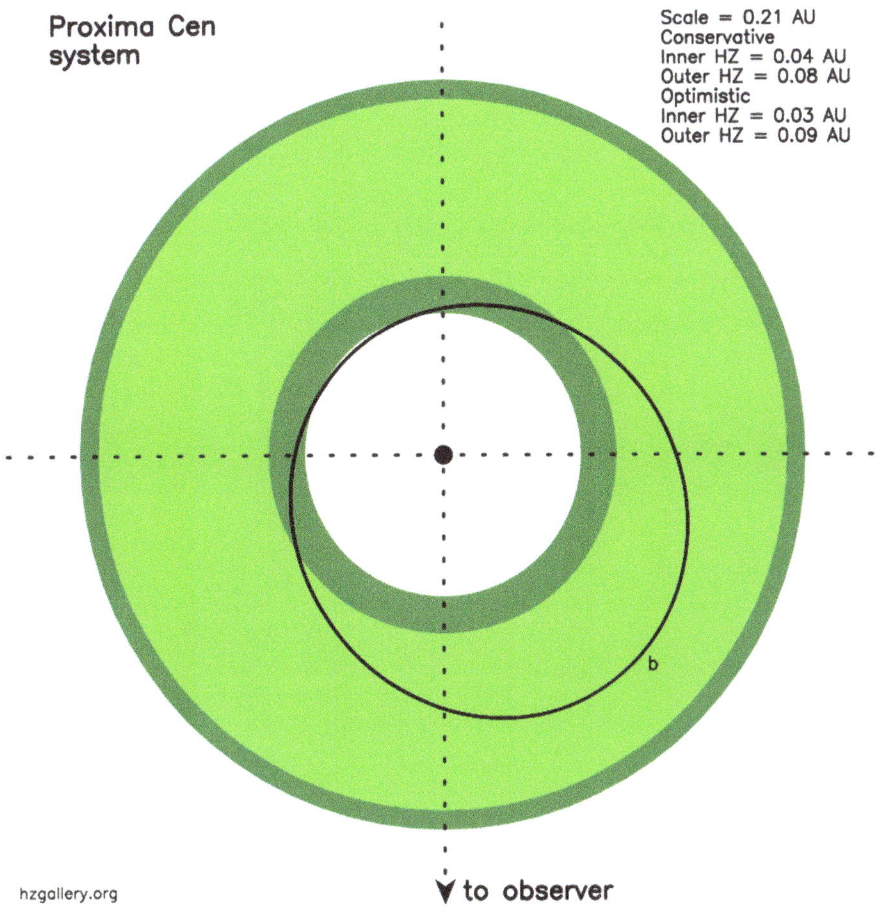

Figure 8.7. A top-down view of the Proxima Centauri system, where the orbit of the known planet is shown and the size along one side (scale) of the figure is indicated in the top-right corner. The host star is located at the center of the dotted cross-hairs and the extent of the HZ is shown in green, where light green is the conservative HZ and dark green is the optimistic extension to the HZ. These data were extracted from the Habitable Zone Gallery (Kane & Gelino 2012).

top-down of the Proxima Centauri system, including the orbit and the extent of the HZ, where light green is the conservative HZ and dark green is the optimistic extension to the HZ. The HZ status of the planet has generated considerable interest in its potential habitability (Ribas et al. 2016; Turbet et al. 2016; Meadows et al. 2018). These studies have included assessments of the UV radiation due to flares on surface conditions (Abrevaya et al. 2020), and three-dimensional climate simulations that consider a range of atmospheric and topographical scenarios (Salazar et al. 2020; Galuzzo et al. 2021). Furthermore, although the planet does not transit, substantial analysis of possible atmospheric characterization opportunities have been estimated based on current and future facilities (Kreidberg & Loeb 2016; Lovis et al. 2017; Luger et al. 2017b). Suffices to say that the research regarding Proxima Centauri b is still in its infancy, and the fact that it will always be our closest, and possibly habitable, exoplanet neighbor ensures that it will remain a target of great interest.

8.3.3 Teegarden's Star

The planetary system of Teegarden's Star is another notable example of HZ terrestrial exoplanets, largely due to its closeness to the Sun (3.8 parsec, or 12.5 light years). The star has a spectral type of M7.0, placing it at the lower end of main sequence stellar masses, and is thus quite faint ($V = 15.4$; Teegarden et al. 2003; Davison et al. 2015). The star was discovered relatively recently (2003) and was noted to have very high tangential velocity relative to the Sun (proper motion; Teegarden et al. 2003). Radial velocity searches for planets orbiting the star were undertaken soon thereafter, with radial velocity variations in the star detected, but

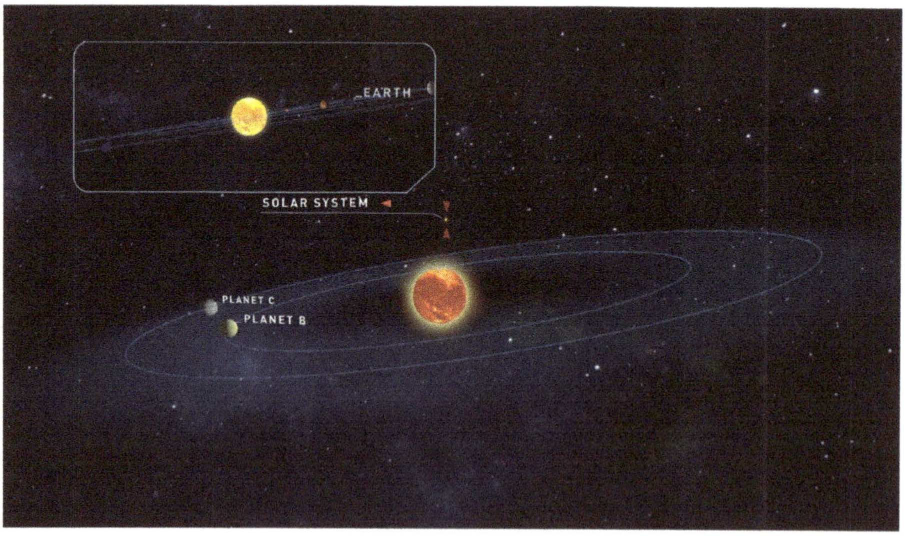

Figure 8.8. Artist's depiction of the two known planets orbiting Teegarden's star, showing the extent of the HZ as a faint green ring, and a size comparison to the solar system. Credit: Reprinted with permission from Zechmeister et al. (2019). Copyright © University of Göttingen, Institute for Astrophysics.

of insufficient significance to claim a planet detection (Barnes et al. 2012). Confirmed detections were announced several years later, finding two planets with orbital periods of 4.91 and 11.4 days, and each having minimum masses of 1.1 $M_{\oplus}$ (Zechmeister et al. 2019). No evidence of planetary transits have been found. Figure 8.8 shows an artist's depiction of the system, highlighting how small the system is in comparison to the solar system and that both planets lie within the HZ. According to the Habitable Zone Gallery, planet c lies in the inner portion of the optimistic HZ and planet c lies near the middle of the conservative HZ. Although claims have been made that the planets could indeed be habitable (Wandel & Tal-Or 2019), such claims require validation from robust atmospheric measurements. The system has difficult prospects for further characterization due to the faintness of the host star, but its relative closeness to our own system raises hopes that new observational techniques and instrumentation may shed further light on this fascinating system.

8.3.4 TOI-700

The TOI-700 system is a notable discovery from the TESS mission, where the acronym "TOI" is short for "TESS Object of Interest". The star is an M2 dwarf, and is located at a distance of 31.1 parsec, or 101.6 light years. The system was found to contain three transiting planets, with radii of 1.037, 2.65, and 1.144 $R_{\oplus}$, and orbits of 9.977, 16.05, and 37.42 days, respectively (Gilbert et al. 2020; Rodriguez et al. 2020). According to the Habitable Zone Gallery, planet d lies near the inner boundary of the conservative HZ, meaning that TOI-700d is a near Earth-size planet in the HZ of its host star. The planetary system discovery was published in a series of three papers that presented the validation of the planetary candidates (Gilbert et al. 2020), confirmation of the planets from observations using the Spitzer Space Telescope (Rodriguez et al. 2020), and an investigation of the potential habitability of planet d using three-dimensional climate simulations (Suissa et al. 2020). Figure 8.9 shows an artist's depiction of planet d and the transit detected through TESS data (top), as well as the overall planetary system, including the host star, orbits, and extent of the HZ (bottom). Studies of the stellar wind and potential atmospheric stripping have found that planet d could maintain long-term atmospheric integrity (Cohen et al. 2020), depending on the strength of the planetary magnetic field (Dong et al. 2020). TOI-700 thus provides an additional source from which to investigate the sustainability of planetary habitability around M dwarf stars. However, the host star has a V magnitude of ~13.2 and so, although brighter than Kepler-186, will still be a challenging target for detailed atmospheric follow-up observations.

8.3.5 TRAPPIST-1

Finally, we return briefly to the TRAPPIST-1 system. We have previously discussed this well-known system in the context of planetary interiors (Section 4.3), atmospheric escape (Section 4.4), HZ examples (Section 6.2), target selection (Section 6.4), and transit probabilities (Section 8.2). Clearly, the TRAPPIST-1 system is an important example of terrestrial planets within the HZ that has captured considerable interest, ranging from planet formation and dynamics around low-mass stars,

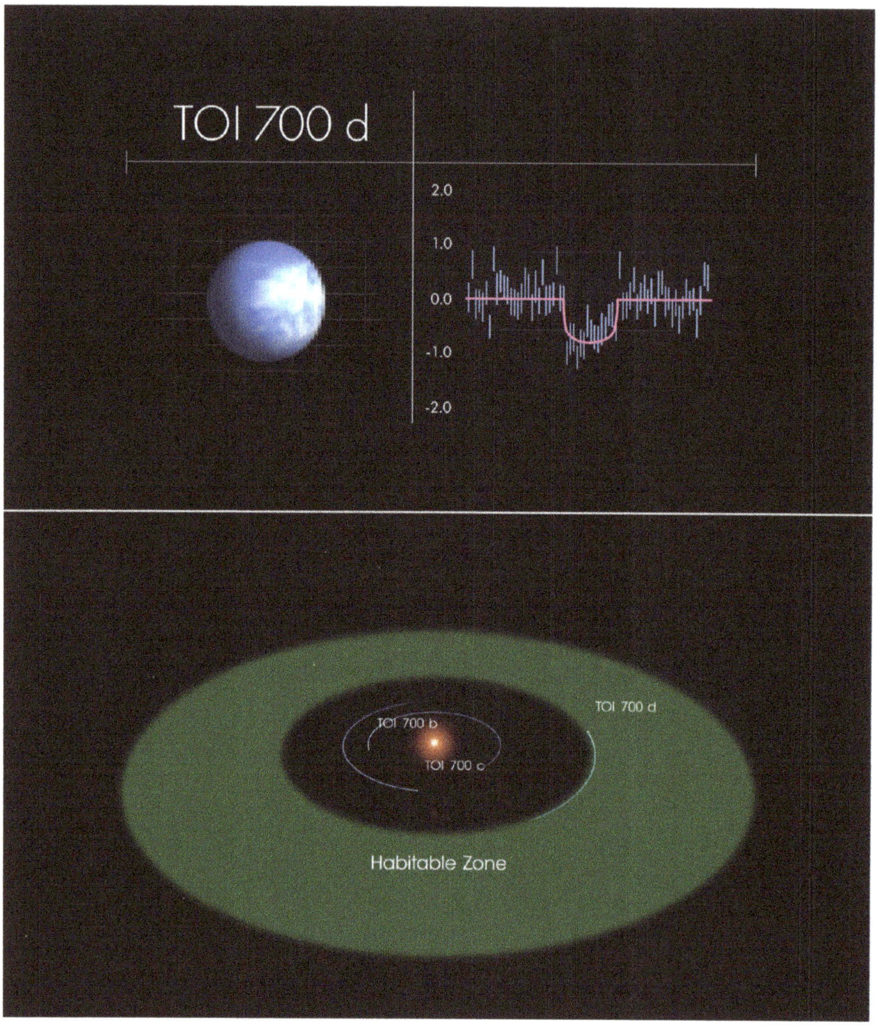

Figure 8.9. Top: artist's depiction of TOI-700d, and the planetary transit as revealed by TESS photometry. Bottom: depiction of the planetary system, showing orbits and the extent of the HZ in green. Credit: Courtesy of NASA's Goddard Space Flight Center/Chris Smith (USRA).

to planetary habitability. Remarkably, this fascinating system was discovered via ground-based transit observations (Gillon et al. 2016, 2017), whereas most other transiting terrestrial HZ planets were acquired through space-based photometry. However, the TRAPPIST-1 system has indeed been the subject of space-based photometry, including the K2 mission (Howell et al. 2014; Luger et al. 2017c), the Hubble Space Telescope (Bourrier et al. 2017b; Zhang et al. 2018), and the Spitzer Space Telescope (Delrez et al. 2018). Furthermore, as noted in Section 6.4, TRAPPIST-1 is expected to become a prime target for James Webb Space Telescope (JWST) transmission spectroscopy observations (Barstow & Irwin 2016; Fauchez et al. 2019; Lustig-Yaeger et al. 2019).

As mentioned in Section 8.2, the star is an ultracool dwarf with a radius of $R_{\star} = 0.12\ R_{\odot}$ and is located at a distance of ~12 parsec. The planet sizes lie in the range 0.752–1.127 $R_{\oplus}$ and thus are all terrestrial in nature. Figure 8.10 shows the planet densities, relative sizes, and incident flux, with solar system terrestrial bodes shown for comparison (Agol et al. 2010). The TRAPPIST-1 planets are all slightly less dense than Earth, possibly due to a higher water content (Unterborn et al. 2018b). The HZ range for both TRAPPIST-1 and the Sun are shown at the top of the figure and, according to the Habitable Zone Gallery (see bottom-right panel of Figure 6.3), planet d lies within the inner optimistic HZ and planets e–g lie within the conservative HZ. This extraordinary grouping of planets within the HZ may be exceptionally common in compact systems around low-mass stars (Ford 2014; Winn & Fabrycky 2015); and possibly even in planetary systems with solar-type stars without a giant planet (Kane et al. 2020). The combination of short periods and orbital resonances in the TRAPPIST-1 system provides further opportunities for rare follow-up observations, such as planet–planet occultations (Luger et al. 2017a) and combined phase variations (Kane et al. 2021). Most importantly, the TRAPPIST-1 system is likely to become a benchmark system in the assessment of terrestrial planet volatile inventory (Unterborn et al. 2018a, 2018b) and atmospheric erosion (Bourrier et al. 2017a). Transmission spectroscopy of the TRAPPIST-1 planets rule out cloud-free, H_2-dominated atmospheres (de Wit et al. 2016, 2018; Wakeford et al. 2019), and a combination of laboratory work and modeling suggests that the atmospheres of these worlds, if they exist, are also unlikely to be H_2-rich and cloudy (Moran et al. 2018). The coming observations of these planets are expected to reveal substantially greater atmospheric details, and may show that terrestrial atmospheres near M dwarfs are largely desiccated. Simply put, the TRAPPIST-1

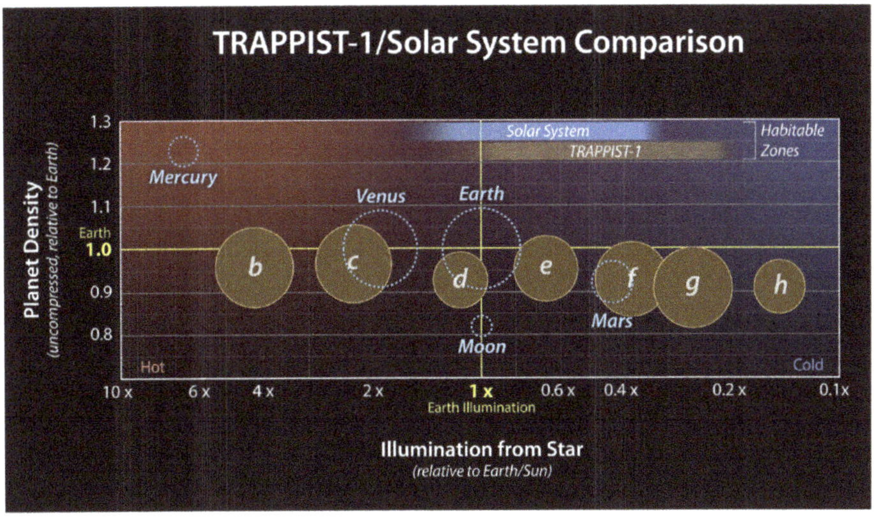

Figure 8.10. Densities of the TRAPPIST-1 planets versus their incident flux (illumination) from the host star (Agol et al. 2010). The solar system terrestrial planets and Earth's Moon are shown for comparison. The extent of the HZ for each system is shown at the top of the plot. Credit: NASA / JPL-Caltech.

may be the most important near-term source of clues regarding the dominant factors that determine planetary habitability.

References

Abrevaya, X. C., Leitzinger, M., Oppezzo, O. J., et al. 2020, MNRAS, 494, L69
Agol, E., & Deck, A. 2016, ApJ, 818, 177
Agol, E., Dorn, C., Grimm, S. L., et al. 2020, PSJ, 2, 1
Akeson, R. L., Chen, X., Ciardi, D., et al. 2013, PASP, 125, 989
Anglada-Escudé, G., Amado, P. J., Berdiñas, Z. M., et al. 2016, Natur, 536, 437
Barclay, T., Quintana, E. V., Raymond, S. N., & Penny, M. T. 2017, ApJ, 841, 86
Barnes, J. R., Jenkins, J. S., Jones, H. R. A., et al. 2012, MNRAS, 424, 591
Barstow, J. K., & Irwin, P. G. J. 2016, MNRAS, 461, L92
Bolmont, E., Raymond, S. N., von Paris, P., et al. 2014, ApJ, 793, 3
Borucki, W. J. 2016, RPPh, 79, 036901
Bourrier, V., de Wit, J., Bolmont, E., et al. 2017a, AJ, 154, 121
Bourrier, V., Ehrenreich, D., Wheatley, P. J., et al. 2017b, A&A, 599, L3
Bowler, P., Blunt, S. C., & Nielsen, E. L. 2020, AJ, 159, 63
Bryson, S., Kunimoto, M., Kopparapu, R. K., et al. 2021, AJ, 161, 36
Burke, C. J. 2008, ApJ, 679, 1566
Catanzarite, J., & Shao, M. 2011, PASP, 123, 171
Clanton, C., & Gaudi, B. S. 2016, ApJ, 819, 125
Cohen, O., Garraffo, C., Moschou, S.-P., et al. 2020, ApJ, 897, 101
Damasso, M., & Del Sordo, F. 2020, MNRAS, 494, 1387
Damasso, M., Del Sordo, F., Anglada-Escudé, G., et al. 2020, SciA, 6, eaax7467
Davenport, J. R. A., Kipping, D. M., Sasselov, D., Matthews, J. M., & Cameron, C. 2016, ApJL, 829, L31
Davison, C. L., White, R. J., Henry, T. J., et al. 2015, AJ, 149, 106
de Wit, J., Wakeford, H. R., Lewis, N. K., et al. 2018, NatAs, 2, 214
de Wit, J., Wakeford, H. R., Gillon, M., et al. 2016, Natur, 537, 69
Delrez, L., Gillon, M., Triaud, A. H. M. J., et al. 2018, MNRAS, 475, 3577
Demory, B.-O., Ehrenreich, D., Queloz, D., et al. 2015, MNRAS, 450, 2043
Desort, M., Lagrange, A. M., Galland, F., Udry, S., & Mayor, M. 2007, A&A, 473, 983
Dong, C., Jin, M., & Lingam, M. 2020, ApJL, 896, L24
Dressing, C. D., & Charbonneau, D. 2013, ApJ, 767, 95
Dressing, C. D., & Charbonneau, D. 2015, ApJ, 807, 45
Dulz, S. D., Plavchan, P., Crepp, J. R., et al. 2020, ApJ, 893, 122
Dumusque, X., Pepe, F., Lovis, C., et al. 2012, Natur, 491, 207
Fauchez, T. J., Turbet, M., Villanueva, G. L., et al. 2019, ApJ, 887, 194
Fischer, D. A., Anglada-Escude, G., Arriagada, P., et al. 2016, PASP, 128, 066001
Ford, E. B. 2014, PNAS, 111, 12616
Foreman-Mackey, D., Hogg, D. W., & Morton, T. D. 2014, ApJ, 795, 64
Fulton, B. J., Petigura, E. A., Howard, A. W., et al. 2017, AJ, 154, 109
Fulton, B. J., Rosenthal, L. J., Hirsch, L. A., et al. 2021, ApJS, 255, 14
Gaidos, E. 2013, ApJ, 770, 90
Galuzzo, D., Cagnazzo, C., Berrilli, F., Fierli, F., & Giovannelli, L. 2021, ApJ, 909, 191
Gilbert, E. A., Barclay, T., Schlieder, J. E., et al. 2020, AJ, 160, 116

Gillon, M., Jehin, E., Lederer, S. M., et al. 2016, Natur, 533, 221
Gillon, M., Triaud, A. H. M. J., Demory, B.-O., et al. 2017, Natur, 542, 456
Goldreich, P., & Soter, S. 1966, Icar, 5, 375
Hartkopf, W. I., Mason, B. D., & Worley, C. E. 2001, AJ, 122, 3472
Hatzes, A. P. 2013, ApJ, 770, 133
Hinkel, N. R., & Kane, S. R. 2013, MNRAS, 432, L36
Howell, S. B., Sobeck, C., Haas, M., et al. 2014, PASP, 126, 398
Jenkins, J. S., Harrington, J., Challener, R. C., et al. 2019, MNRAS, 487, 268
Jurić, M., & Tremaine, S. 2008, ApJ, 686, 603
Kamper, K. W., & Wesselink, A. J. 1978, AJ, 83, 1653
Kane, S. R. 2018, ApJL, 861, L21
Kane, S. R., Ciardi, D. R., Gelino, D. M., & von Braun, K. 2012, MNRAS, 425, 757
Kane, S. R., & Gelino, D. M. 2012, PASP, 124, 323
Kane, S. R., Gelino, D. M., & Turnbull, M. C. 2017, AJ, 153, 52
Kane, S. R., & von Braun, K. 2008, ApJ, 689, 492
Kane, S. R., Jansen, T., Fauchez, T., Selsis, F., & Ceja, A. Y. 2021, AJ, 161, 53
Kane, S. R., Turnbull, M. C., Fulton, B. J., et al. 2020, AJ, 160, 81
Kervella, P., Thévenin, F., & Lovis, C. 2017, A&A, 598, L7
Kipping, D. M. 2014, MNRAS, 444, 2263
Kipping, D. M., Cameron, C., Hartman, J. D., et al. 2017, AJ, 153, 93
Kopparapu, R. K. 2013, ApJ, 767, L8
Kreidberg, L., & Loeb, A. 2016, ApJL, 832, L12
Krishnamurthy, A., Knapp, M., Günther, M. N., et al. 2021, AJ, 161, 275
Kunimoto, M., & Matthews, J. M. 2020, AJ, 159, 248
Laskar, J., Boué, G., & Correia, A. C. M. 2012, A&A, 538, A105
Lissauer, J. J., Ragozzine, D., Fabrycky, D. C., et al. 2011, ApJS, 197, 8
Lissauer, J. L., Marcy, G. W., Rowe, J. F., et al. 2012, ApJ, 750, 112
Liu, H.-G., Jiang, P., Huang, X., et al. 2018, AJ, 155, 12
Lopez, E. D., & Fortney, J. J. 2013, ApJ, 776, 2
Lovis, C., Snellen, I., Mouillet, D., et al. 2017, A&A, 599, A16
Luger, R., Lustig-Yaeger, J., & Agol, E. 2017a, ApJ, 851, 94
Luger, R., Lustig-Yaeger, J., Fleming, D. P., et al. 2017b, ApJ, 837, 63
Luger, R., Sestovic, M., Kruse, E., et al. 2017c, NatAs, 1, 0129
Lustig-Yaeger, J., Meadows, V. S., & Lincowski, A. P. 2019, AJ, 158, 27
MacGregor, M. A., Weinberger, A. J., Wilner, D. J., Kowalski, A. F., & Cranmer, S. R. 2018, ApJL, 855, L2
Matthews, R., & Gilmore, G. 1993, MNRAS, 261, L5
Meadows, V. S., Arney, G. N., Schwieterman, E. W., et al. 2018, AsBio, 18, 133
Moorhead, A. V., Ford, E. B., Morehead, R. C., et al. 2011, ApJS, 197, 1
Moran, S. E., Hörst, S. H., Batalha, N. E., Lewis, N. K., & Wakeford, H. R. 2018, AJ, 156, 252
Nielsen, E. L., De Rosa, R. J., Macintosh, B., et al. 2019, AJ, 158, 13
Owen, J. E., & Wu, Y. 2013, ApJ, 775, 105
Pepe, F., Cristiani, S., Rebolo, R., et al. 2021, A&A, 645, A96
Petersburg, R. R., Ong, J. M. J., Zhao, L. L., et al. 2020, AJ, 159, 187
Quintana, E. V., Barclay, T., Raymond, S. N., et al. 2014, Sci, 344, 277
Ramos, X. S., Charalambous, C., Benítez-Llambay, P., & Beaugé, C. 2017, A&A, 602, A101

Ribas, I., Bolmont, E., Selsis, F., et al. 2016, A&A, 596, A111
Robertson, P., Anderson, T., Stefansson, G., et al. 2019, JATIS, 5, 015003
Robertson, P., Bender, C., Mahadevan, S., Roy, A., & Ramsey, L. W. 2016, ApJ, 832, 112
Rodriguez, J. E., Vanderburg, A., Zieba, S., et al. 2020, AJ, 160, 117
Rowe, J. F., Bryson, S. T., Marcy, G. W., et al. 2014, ApJ, 784, 45
Saar, S. H., & Donahue, R. A. 1997, ApJ, 485, 319
Salazar, A. M., Olson, S. L., Komacek, T. D., Stephens, H., & Abbot, D. S. 2020, ApJL, 896, L16
Shen, Y., & Turner, E. L. 2008, ApJ, 685, 553
Suissa, G., Wolf, E. T., Kopparapu, R. K., et al. 2020, AJ, 160, 118
Teegarden, B. J., Pravdo, S. H., Hicks, M., et al. 2003, ApJL, 589, L51
Torres, G., Kipping, D. M., Fressin, F., et al. 2015, ApJ, 800, 99
Turbet, M., Leconte, J., Selsis, F., et al. 2016, A&A, 596, A112
Unterborn, C. T., Dismukes, E. E., & Panero, W. R. 2016, ApJ, 819, 32
Unterborn, C. T., Hinkel, N. R., & Desch, S. J. 2018a, RNAAS, 2, 116
Unterborn, C. T., Desch, S. J., Hinkel, N. R., & Lorenzo, A. 2018b, NatAs, 2, 297
Van Eylen, V., & Albrecht, S. 2015, ApJ, 808, 126
Wakeford, H. R., Lewis, N. K., Fowler, J., et al. 2019, AJ, 157, 11
Walsh, K. J., & Morbidelli, A. 2011, A&A, 526, A126
Wandel, A., & Tal-Or, L. 2019, ApJL, 880, L21
Winn, J. N., & Fabrycky, D. C. 2015, ARA&A, 53, 409
Wittenmyer, R. A., Butler, R. P., Tinney, C. G., et al. 2016, ApJ, 819, 28
Wittenmyer, R. A., Butler, R. P., Horner, J., et al. 2020, MNRAS, 491, 5248
Wright, J. T. 2005, PASP, 117, 657
Zechmeister, M., Dreizler, S., Ribas, I., et al. 2019, A&A, 627, A49
Zechmeister, M., Kürster, M., Endl, M., et al. 2013, A&A, 552, A78
Zhang, Z., Zhou, Y., Rackham, B. V., & Apai, D. 2018, AJ, 156, 178

AAS | IOP Astronomy

Planetary Habitability

Stephen R Kane

Chapter 9

The Next Steps[1]

9.1 A Strategic Imperative

The past several decades have witnessed enormous strides in planetary science, exoplanet detection, and exoplanet characterization. In particular, the exoplanet field is on the brink of terrestrial exoplanet atmospheric observations that will be placed within the context of solar system data. Transmission spectra have been obtained for a suite of hot Jupiters, revealing a diversity of atmospheric compositions, clear or cloudy atmospheres, and temperature profiles (Sing et al. 2016). Observations of exo-Neptunes have revealed flat to water absorption dominated spectra that show possible trends in composition with size (Crossfield & Kreidberg 2017). Of highest importance are the advances regarding observational constraints on terrestrial worlds that suggest these planets may have high-molecular-weight atmospheres like the terrestrial planets in our own system, although the data are also consistent with no atmospheres at all. James Webb Space Telescope (JWST) observations of the TRAPPIST-1 system and other nearby terrestrial worlds will have the capability to detect the presence and composition of atmospheres (Morley et al. 2017; Lincowski et al. 2018; Lustig-Yaeger et al. 2019b; Wunderlich et al. 2019) potentially revealing past processes, such as atmosphere and ocean-loss (Lincowski et al. 2018, 2019; Lustig-Yaeger et al. 2019a). These observations may also provide our first opportunity to search for signs of life, such as CH_4 in combination with other biosignatures, in the atmosphere of a terrestrial exoplanet (Krissansen-Totton et al. 2018; Wunderlich et al. 2019) and complement observations from the ground with extremely large telescopes that will search for O_2 using high resolution spectroscopy (Lovis et al. 2017; López-Morales et al. 2019). Shown in Figure 9.1 is an example synthetic spectrum, covering a broad wavelength range, that represents an Earth radiance spectrum when viewed at quadrature (half illumination; Schwieterman et al. 2018). Notice that the top panel, which covers

[1] Parts of this chapter have been adapted from Kane et al. (2021).

doi:10.1088/2514-3433/ac2aa1ch9

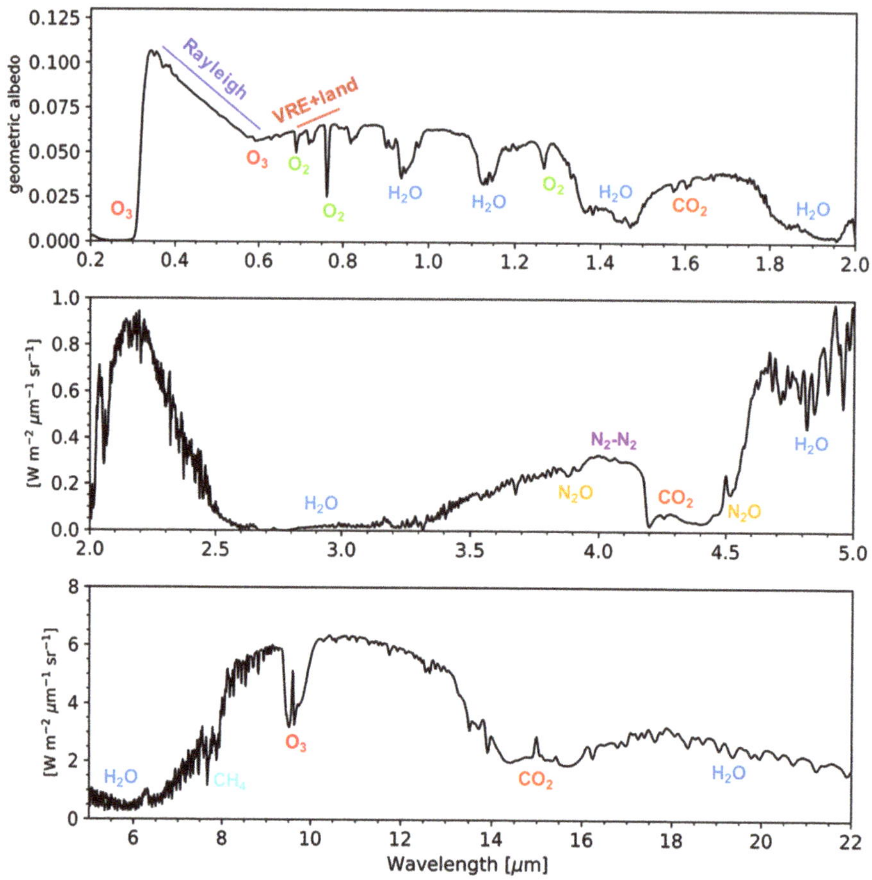

Figure 9.1. A synthetic broadband Earth radiance spectrum at quadrature phase (half illumination), showing the location of the strong absorption features and Rayleigh scattering. Credit: Schwieterman et al. (2018).

the visible wavelength range, is shown in terms of the geometric albedo. Strong absorption features from O_2, O_3, H_2O, CO_2, N_2O, and CH_4 are labeled, in addition to Rayleigh scattering and the location of the vegetation red edge (VRE), where chlorophyll dramatically changes the absorption properties of vegetation (Seager et al. 2005; Tinetti et al. 2006; O'Malley-James & Kaltenegger 2018). Detection of such biosignatures has the potential to provide essential guidance for which planets are indeed habitable, and thus whose overall planetary properties may then be assessed in their role toward enabling temperate surface conditions accordingly.

Recall from Chapter 2 that the primary goal of planetary habitability studies is to catalog all of the possible habitability factors, then to rank them according to their influence on surface conditions. Some factors will be primary contributors to habitability, while others will have second or third order effects. Not only is this an extremely difficult undertaking, but is also likely case specific in a way that is not yet obvious. Even so, the task of ranking habitability factors is a vast interdisciplinary effort that benefits all of the related topics, including the future and sustainability of Earth's environment. The planetary habitability story can therefore be considered a strategic imperative in the sense that it addresses age-old questions regarding life in the Universe in addition to ensuring that the survival of our species overlaps with answering those questions.

The strategic approach to issues regarding planetary habitability have been addressed via the NASA Strategic Goals.[2] For example, the "Science 2020–2024: A Vision for Scientific Excellence" contains the following quotes which have been drawn from the description of the three major disciplines.

1. Earth Science: "NASA Earth Science unlocks the mysteries of our planet, exploring, discovering, and responding to the need to understand our planet's interconnected systems, from a global scale to minute processes."
2. Heliophysics: "The study of the coupled solar-terrestrial system can also teach us more about the habitability of planets in other stellar systems throughout the universe."
3. Planetary Science: "Through the observation and discovery of complex planetary worlds and objects, we seek to understand our solar system and the distribution of life within it. The NASA Planetary Science strategic objective is to advance scientific knowledge of the origin and history of the solar system, the potential for life elsewhere, and the hazards and resources present as humans explore space."

Though broad, these areas of research encompass many of the aspects required for understanding planetary habitability and underlines the need to approach these topics in parallel.

The first objective of the "NASA 2018 Strategic Plan" is as follows: "Strategic Objective 1.1: Understand the Sun, Earth, Solar System, and Universe." The description of this goal contains the following excerpt, where I have highlighted the final sentence for emphasis: "The search for life in the solar system and beyond is guided by the ability to understand how life originated on Earth and by the quest to find habitable environments outside of Earth. To improve the knowledge of environmental requirements for habitability, NASA will develop tools for detecting life, develop tools for determining the relative habitability of present or ancient environments, and explore analog environments on Earth. This will facilitate target selection for further robotic, and ultimately human, exploration. Observations from SMD's astrophysics missions have made it clear that habitable planets exist around stars other than the Sun and that such planets are plentiful. Improving techniques and ideas for discovering and characterizing habitable and/or inhabited environments on these planets, coupled with an understanding of the potential false positives for habitability or life, will enable prioritization of exoplanets for targeted follow-up observations. This will help to push frontiers in the coming decades of discovery and enable the search for signs of life on worlds that may be capable of harboring life, both within our own solar system and within the Galaxy. *NASA's*

[2] https://science.nasa.gov/about-us/science-strategy

strategy relies on applying the lessons learned about the origin, evolution, and distribution of life on Earth to other bodies in our solar system and beyond." The reason that the last sentence is emphasized is because it stresses, once again, the need for a collaborative effort that builds upon the *in situ* data acquired locally. Both the exoplanet and planetary science communities are moving toward a more systems and processes-based approach to understanding planet formation, evolution, habitability and biosignatures. These approaches require the synthesis of observation, theory, and laboratory work from multiple disciplines, and it is very clear that the two communities can benefit greatly from the knowledge and perspectives provided by both of their fields.

9.2 Local Lessons

Although planetary science and exoplanetary science are highly complementary fields, they evolved on separate pathways, largely for historical reasons. Exoplanet detection has primarily been a task of stellar characterization by stellar astronomers, from which the presence of a companion of planetary size/mass may be inferred, whereas planetary science has focused on specific worlds within the solar system. The discoveries of numerous terrestrial exoplanets has provoked further discourse between the disciplines with a common objective of leveraging solar system data toward a deeper understanding of exoplanet observables (Kane et al. 2021). Exoplanetary science aims to achieve reliable pathways towards accurate characterization of exoplanet atmospheres, surfaces, and interiors. However, correct interpretation of all these data relies upon models of planetary processes that are best developed and validated using *in situ* or remote-sensing observations of solar system bodies (Fujii et al. 2014). Similarly, observations of solar system bodies, even very fundamental ones like phase dependent photometry of the Jovian planets (Mayorga et al. 2016), or simulated transmission observations of Titan (Robinson et al. 2014) or the Earth (Macdonald & Cowan 2019) can help inform planning and interpretation of exoplanet observations, and help us train predictive (forward) and retrieval (inverse) models for exoplanets.

Planetary science thus forms the foundation, both in terms of data and models, from which exoplanet observables may be interpreted. In turn, exoplanet observables provide vast numbers of exoplanets from which demographic studies can inform the studies of planetary system formation and evolution in general (Clanton & Gaudi 2016; Barclay et al. 2017; Nielsen et al. 2019; see Section 8.1). Measurements from planetary science and exoplanet observables feed into the inferred properties of planetary bodies generally and, in particular, the potential surface conditions of a terrestrial exoplanet that may have temperate surface conditions. Models that are well-validated on solar system bodies, especially Earth, will be particularly critical for the difficult task of inferring the presence of life on an exoplanet from possible observed biosignatures. Indeed, this may be the most challenging task faced by planetary scientists in the coming decades when spectral observations of potentially habitable terrestrial planets become possible, and inter-disciplinary collaborations of scientists will be essential to its success.

The pathway forward therefore lies in identifying the key measurables from solar system bodies, through *in situ* observations with spacecraft missions and those acquired on and in orbit of Earth (Robinson et al. 2011; Jiang et al. 2018) needed to correctly interpret exoplanet observables and infer their properties. Examples of key planetary science questions relevant to exoplanetary science include (but are by no means limited to):

- What effect, if any, has the proximity of Mercury to the host star had on the planet's abundances of moderately volatile elements such as C and S (Vander Kaaden et al. 2020)?
- Where did Venus' water go, and what processes are most important for O_2 loss from terrestrial planet atmospheres? Was hydrogen loss and abiotic oxygen production prevalent, or did surface hydration dominate (Kasting et al. 1984; Watson et al. 1984; Lichtenegger et al. 2016)?
- Did Venus have a habitable period (Way et al. 2016; Way & Del Genio 2020)? That is, did Venus ever cool after formation (Hamano et al. 2013)? If Venus had a habitable period, how long did it last—and when did it end?
- What is the role of volatiles (e.g., liquid water) in Earth's continental plate subduction and carbon cycle (Regenauer-Lieb et al. 2001; Bercovici 2003; van der Lee et al. 2008) and how critical is this process for the sustained habitability of terrestrial exoplanets (Valencia et al. 2007; Lammer et al. 2009; Lena & Breuer 2014)?
- To what extent was early Mars really "warm and wet" in contrast to the cold and dry planet we see today (Ramirez & Craddock 2018; Grau Galofre 2020) and what does Mars teach us about the lower size limit for sustained planetary habitability (Ehlmann et al. 2016)?

In the near-term, the most critical data needed from planetary science are atmospheric measurements that can constrain composition, chemistry, dynamics and evolutionary history, particularly for poorly understood atmospheres such as those of Venus (Kane et al. 2019), Titan (Checlair et al. 2016), and the ice giants (Wakeford & Dalba 2020). Beyond modeling the nature of exoplanet atmospheres and inferring possible surface conditions lies the complex task of modeling how the interior and surface have previously, and are currently, interacting with the atmosphere. In particular, a major challenge lies in distinguishing between biotic and abiotic processes that yield gases of biological significance (e.g., CH_4, CO_2, H_2O), the signatures of which can be detected from studying atmospheric abundances (Meadows 2017; Fujii et al. 2018; Harman et al. 2018; Wogan & Catling 2020; see Figure 9.1).

Fortunately, there is an active portfolio of current and planned future planetary science missions, summarized in Figure 9.2, which is current as of 2021 July 1. The design of many recent solar system exploration proposals and selected missions have incorporated science goals that specifically benefit anticipated studies of exoplanets. Furthermore, laboratory experiments are being conducted that provide the framework for interpreting molecular absorption in exoplanet atmospheres (Tennyson & Yurchenko 2017; Hörst et al. 2018; He et al. 2020; Moran et al. 2020). Therefore, it is expected that the continuing reliance of exoplanetary science on solar system

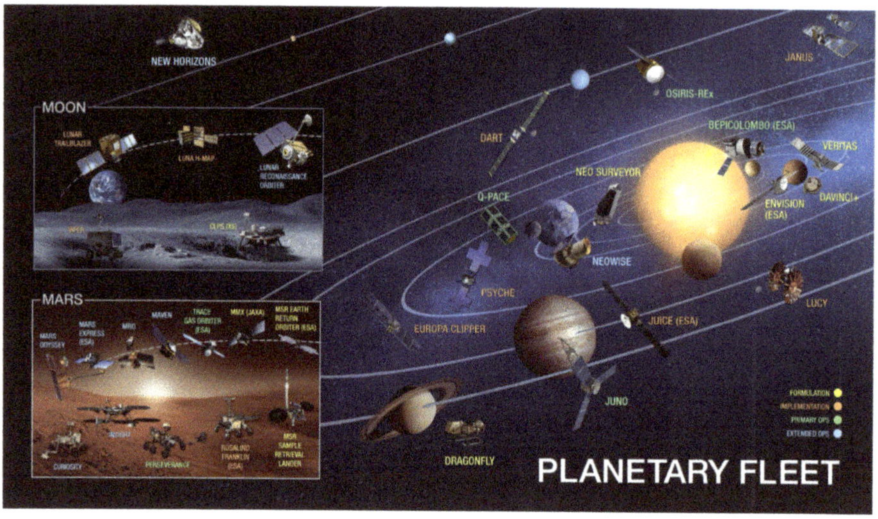

Figure 9.2. NASA current and future planetary science missions, as of 2021 July 1. Credit: NASA.

exploration will provide enormous benefits for both fields of research as the vast data provided by the plethora of exoplanets answers significant questions regarding the context and uniqueness of the solar system and, ultimately, the prevalence of life in the universe.

9.3 Unlocking Achievements

The full range of parameters that influence planetary habitability, described in Section 1.1, is both extensive and intimidating. Although many exoplanets are known, the current detection techniques that have uncovered the bulk of the exoplanet inventory provide very limited information regarding their true nature (see Section 8.2), and many of the parameters described in Figure 2.1 are presently unmeasurable. Meanwhile, significant questions remain right here in our own planetary system where, for example, it is still unclear how, why, or when the evolution of Venus and Earth diverged. There is much to be done.

Both NASA and ESA are developing plans for future facilities that create a pathway toward addressing the current gaps in our knowledge of exoplanets, summarized in Figure 9.3. For example, the NASA mission pathway beyond the TESS mission is one primarily of characterizing the known exoplanets and testing techniques that will enable us to unlock the ability to measure parameters that are critically important for planetary habitability. The Nancy Grace Roman Space Telescope (formerly WFIRST) equipped with a coronagraph will serve as a path-finder and technology demonstration for achieving exoplanet direct imaging goals (Kasdin et al. 2020). Mission concepts, such as LUVOIR (The LUVOIR Team 2019) and the Habitable Exoplanet Observatory (HabEx) Mission Concept Study (Gaudi et al. 2001) will be crucial next steps toward a full terrestrial characterization (Kopparapu et al. 2018). These missions will employ a direct imaging approach that avoids the crippling requirement of orbital alignment

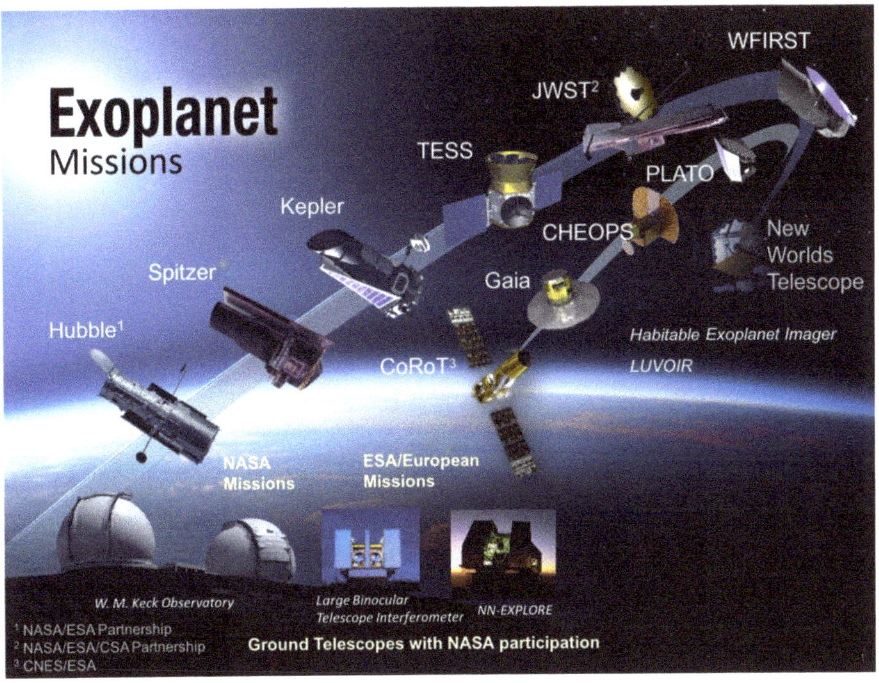

Figure 9.3. The NASA/ESA Exoplanet Mission Roadmap, showing past, present, and future exoplanet missions from space agencies and observatories around the world. Credit: Courtesy of NASA/JPL-Caltech.

when surveying the nearest stars. Furthermore, these missions will provide the means to measure rotation (Guzewich et al. 2020), obliquity (Kane & Torres 2017), seasonal variations (Olson et al. 2018), and many other intrinsic planetary properties that are currently only accessible for solar system objects. An example of simulated data for five predicted exoplanet discoveries from the HabEx mission is shown in Figure 9.4. The simulated spectra assume a wavelength coverage of 0.2–1.8 μm that will address many of the important biosignature features shown in Figure 9.1. These missions will provide "family portraits" for a range of planetary types that are at many different stages of planetary evolution.

Consider that, although Figures 9.2 and 9.3 present two different mission portfolios, the topic of planetary habitability merges them together into one coherent mission strategy. Given the progress that has been made, we can not afford for the exoplanet and planetary science communities to retreat back to the insular environment of their respective disciplines. More than ever, now is the time to invest in inter-disciplinary science to continue progressing the astrobiology narrative. One may wonder if we really need to unlock all of the stellar and planetary parameters described if biosignatures are indeed detected, which would mean that the planet is, by definition, habitable. The answer is a resounding yes because, without understanding the planet and its local environment, even if we know it is habitable, we won't know how or why, nor will we be able to predict habitability elsewhere. Chapter 2 discusses the primary goal of producing an

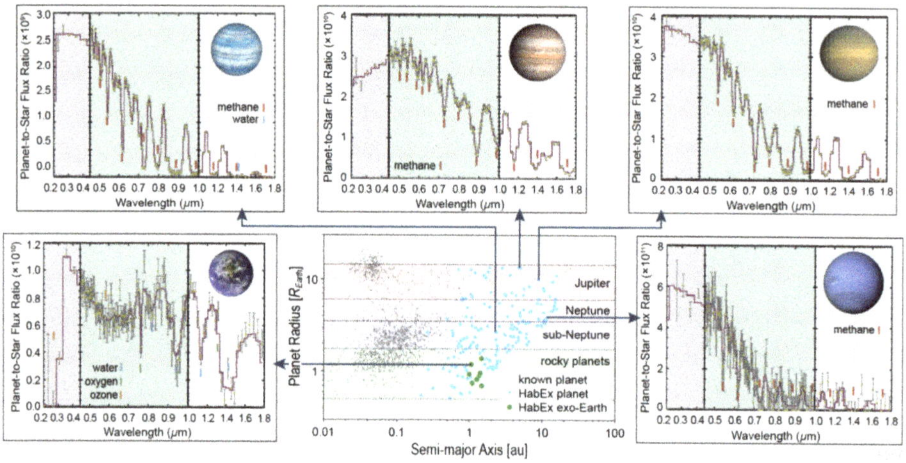

Figure 9.4. A sample of simulated spectra of planets that are predicted to be discovered and characterized by the HabEx mission. HabEx will be able to achieve spectral coverage of 0.2–1.8 μ m for most planets when coupled with a starshade. Credit: Reprinted with permission from Gaudi et al. (2001). Courtesy of NASA/JPL-Caltech.

exhaustive list of factors (intrinsic and external) that influence the sustainability of temperate planetary surface conditions, then ranking that list in order of the amplitude of their affects. As we have seen, such a task is non-trivial to say the least, largely due to the inter-dependency of many factors and the difficulty in their measurements. Despite these challenges, the task must be addressed if planetary data are to be useful for properly assessing the habitability of their progenitor.

Throughout this book, I have referred to planetary habitability as being extremely complex, often controversial, and sometimes even divisive. However, it is the multi-faceted and inter-disciplinary nature of the subject that creates this intense discussion, and this diversity of experience will ultimately become its greatest unifying strength. The journey is just beginning.

References

Barclay, T., Quintana, E. V., Raymond, S. N., & Penny, M. T. 2017, ApJ, 841, 86

Bercovici, D. 2003, E&PSL, 205, 107

Checlair, J., McKay, C. P., & Imanaka, H. 2016, P&SS, 129, 1

Clanton, C., & Gaudi, B. S. 2016, ApJ, 819, 125

Crossfield, I. J. M., & Kreidberg, L. 2017, AJ, 154, 261

Ehlmann, B. L., Anderson, F. S., Andrews-Hanna, J., et al. 2016, JGRE, 121, 1927

Fujii, Y., Angerhausen, D., Deitrick, R., et al. 2018, AsBio, 18, 739

Fujii, Y., Kimura, J., Dohm, J., & Ohakte, M., et al. 2014, AsBio, 14, 753

Gaudi, B. S., Seager, S., Mennesson, B., et al. 2001, arXiv:2001.06683

Grau Galofre, A., Mark Jellinek, A., & Osinski, G. R. 2020, NatGe, 13, 663

Guzewich, S. D., Lustig-Yaeger, J., Davis, C. E., et al. 2020, ApJ, 893, 140

Hamano, K., Abe, Y., & Genda, H. 2013, Natur, 497, 607

Harman, C. E., Felton, R., Hu, R., et al. 2018, ApJ, 866, 56

He, C., Hörst, S. M., Lewis, N. K., et al. 2020, PSJ, 1, 51
Hörst, S. M., He, C., Lewis, N. K., et al. 2018, NatAs, 2, 303
Jiang, J. H., Zhai, A. J., Herman, J., et al. 2018, AJ, 156, 26
Kane, S. R., & Torres, S. M. 2017, AJ, 154, 204
Kane, S. R., Arney, G., Crisp, D., et al. 2019, JGRE, 124, 2015
Kane, S. R., Arney, G. N., Byrne, P. K., et al. 2021, JGRE, 126, e06643
Kasdin, N. J., Bailey, V. P., Mennesson, B., et al. 2020, Proc. SPIE, 11443, 114431U
Kasting, J. F., Pollack, J. B., & Ackerman, T. P. 1984, Icar, 57, 335
Kopparapu, R. K., Hébrard, E., Belikov, R., et al. 2018, ApJ, 856, 122
Krissansen-Totton, J., Garland, R., Irwin, P., & Catling, D. C. 2018, AJ, 156, 114
Lammer, H., Bredehöft, J. H., Coustenis, A., et al. 2009, A&ARv, 17, 181
Lichtenegger, H. I. M., Kislyakova, K. G., Odert, P., et al. 2016, JGRA, 121, 4718
Lincowski, A. P., Lustig-Yaeger, J., & Meadows, V. S. 2019, AJ, 158, 26
Lincowski, A. P., Meadows, V. S., Crisp, D., et al. 2018, ApJ, 867, 76
López-Morales, M., Ben-Ami, S., Gonzalez-Abad, G., et al. 2019, AJ, 158, 24
Lovis, C., Snellen, I., Mouillet, D., et al. 2017, A&A, 599, A16
Lustig-Yaeger, J., Meadows, V. S., & Lincowski, A. P. 2019a, ApJL, 887, L11
Lustig-Yaeger, J., Meadows, V. S., & Lincowski, A. P. 2019b, AJ, 158, 27
Macdonald, E. J. R., & Cowan, N. B. 2019, MNRAS, 489, 196
Mayorga, L. C., Jackiewicz, J., Rages, K., et al. 2016, AJ, 152, 209
Meadows, V. S. 2017, AsBio, 17, 1022
Moran, S. E., Hörst, S. M., Vuitton, V., et al. 2020, PSJ, 1, 17
Morley, C. V., Kreidberg, L., Rustamkulov, Z., Robinson, T., & Fortney, J. J. 2017, ApJ, 850, 121
Nielsen, E. L., De Rosa, R. J., Macintosh, B., et al. 2019, AJ, 158, 13
Noack, L., & Breuer, D. 2014, P&SS, 98, 41
Olson, S. L., Schwieterman, E. W., Reinhard, C. T., et al. 2018, ApJL, 858, L14
O'Malley-James, J. T., & Kaltenegger, L. 2018, AsBio, 18, 1123
Ramirez, R. M., & Craddock, R. A. 2018, NatGe, 11, 230
Regenauer-Lieb, K., Yuen, D. A., & Branlund, J. 2001, Sci, 294, 578
Robinson, T. D., Maltagliati, L., Marley, M. S., & Fortney, J. J. 2014, PNAS, 111, 9042
Robinson, T. D., Meadows, V. S., Crisp, D., et al. 2011, AsBio, 11, 393
Schwieterman, E. W., Kiang, N. Y., Parenteau, M. N., et al. 2018, AsBio, 18, 663
Seager, S., Turner, E. L., Schafer, J., & Ford, E. B. 2005, AsBio, 5, 372
Sing, D. K., Fortney, J. J., Nikolov, N., et al. 2016, Natur, 529, 59
Tennyson, J., & Yurchenko, S. J. 2017, MolAs, 8, 1
The LUVOIR Team 2019, arXiv:1912.06219
Tinetti, G., Rashby, S., & Yung, Y. L. 2006, ApJL, 644, L129
Valencia, D., O'Connell, R. J., & Dimitar, D. D. 2007, ApJL, 670, L45
van der Lee, S., Regenauer-Lieb, K., & Yuen, D. A. 2008, E&PSL, 273, 15
Vander Kaaden, K. E., McCubbin, F. M., Turner, A. A., & Kent, D. A. 2020, JGRE, 125, e06239
Wakeford, H. R., & Dalba, P. A. 2020, RSPTA, 378, 20200054
Watson, A. J., Donahue, T. M., & Kuhn, W. R. 1984, E&PSL, 68, 1
Way, M. J., & Del Genio, A. D. 2020, JGRE, 125, e06276
Way, M. J., Del Genio, A. D., Kiang, N. Y., et al. 2016, GeoRL, 43, 8376
Wogan, N. F., & Catling, D. C. 2020, ApJ, 892, 127
Wunderlich, F., Godolt, M., Grenfell, J. L., et al. 2019, A&A, 624, A49

www.ingramcontent.com/pod-product-compliance
Ingram Content Group UK Ltd.
Pitfield, Milton Keynes, MK11 3LW, UK
UKHW061857100926
8556IPUK00008B/909

9 780750 321181